Springer Tracts in Advanced Robotics

Volume 37

Editors: Bruno Siciliano · Oussama Khatib · Frans Groen

Aníbal Ollero and Iván Maza (Eds.)

Multiple Heterogeneous Unmanned Aerial Vehicles

Springer

Professor Bruno Siciliano, Dipartimento di Informatica e Sistemistica, Università di Napoli Federico II, Via Claudio 21, 80125 Napoli, Italy, E-mail: siciliano@unina.it

Professor Oussama Khatib, Robotics Laboratory, Department of Computer Science, Stanford University, Stanford, CA 94305-9010, USA, E-mail: khatib@cs.stanford.edu

Professor Frans Groen, Department of Computer Science, Universiteit van Amsterdam, Kruislaan 403, 1098 SJ Amsterdam, The Netherlands, E-mail: groen@science.uva.nl

Editors

Professor Aníbal Ollero
Head of the Robotics, Vision and Control Group,
Escuela Superior de Ingenieros,
The University of Seville
Camino de los Descubrimientos, s/n
41092 Seville
Spain
E-mail: aollero@cartuja.us.es

Professor Iván Maza
Escuela Superior de Ingenieros
The University of Seville
Camino de los Descubrimientos, s/n
41092 Seville
Spain
E-mail: imaza@cartuja.us.es

ISSN print edition: 1610-7438
ISSN electronic edition: 1610-742X

ISBN-10 3-540-73957-2 Springer Berlin Heidelberg New York
ISBN-13 978-3-642-44482-1 Springer Berlin Heidelberg New York

Springer is a part of Springer Science+Business Media
springer.com

Softcover re-print of the Hardcover 1st edition 2007

Typesetting: Digital data supplied by editor.
Data-conversion and production: SPS, Chennai, India
Printed on acid-free paper SPIN: 11854449 89/SPS 5 4 3 2 1 0

This book is dedicated to our respective families for their continuous love and support

Foreword

At the dawn of the new millennium, robotics is undergoing a major transformation in scope and dimension. From a largely dominant industrial focus, robotics is rapidly expanding into the challenges of unstructured environments. Interacting with, assisting, serving, and exploring with humans, the emerging robots will increasingly touch people and their lives.

The goal of the new series of *Springer Tracts in Advanced Robotics (STAR)* is to bring, in a timely fashion, the latest advances and developments in robotics on the basis of their significance and quality. It is our hope that the wider dissemination of research developments will stimulate more exchanges and collaborations among the research community and contribute to further advancement of this rapidly growing field.

The edited volume by Aníbal Ollero and Iván Maza is focused on aerial robotics, a broad area of research in field robotics where aerial robots can overcome the locomotion limitations of ground robots. Unmanned Aerial Vehicles (UAVs) are self-propelled air vehicles that are either remotely controlled by a human operator or are capable of conducting autonomous operations, including search and rescue, security, traffic surveillance, civil engineering, etc.

The book addresses the current state of technology, the existing problems and potentialities of systems consisting of multiple UAVs which are heterogeneous in view of the different characteristics of the aerial vehicles, the different on-board payloads, and the different on-board information processing capabilities. The material is organised in nine chapters by well-recognised authors in the field, covering the core methods and technologies in multi-UAV systems, including prototypes such as autonomous helicopters and airships, as well as discussing future research, development and applications.

As the first focused STAR volume in the growing area of aerial robotics, this title constitutes a fine addition to the series!

Naples, Italy,
May 2007

Bruno Siciliano
STAR Editor

Preface

Aerial robotics is a growing field with many application possibilities from local inspection of utilities and instrumentation deployment in inaccessible sites to global environment monitoring, and including Search and Rescue, Security, Disaster Management, Traffic Surveillance, Civil Engineering and many other field robotics applications. In all these applications aerial robots can overcome the locomotion limitations of ground robots to navigate in rough terrains or to pass over obstacles that cannot be avoided.

Aerial robots can be considered as an evolution of the Unmanned Aerial Vehicles (UAVs). UAVs are self propelled air vehicles that are either remotely controlled by a human operator or are capable of conducting autonomous operations. During the last decades significant efforts have been devoted to increase the flight endurance, flight range and payload of UAVs. Today, UAVs with several thousands of Kilometres of flight range, more than 24 hours of flight endurance and more than 1000 Kilograms payload are in operation.

Furthermore, UAV technology has also evolved to increase the on-board computation and communication capabilities. The development of new navigation sensors, actuators, embedded control and communication systems and the tendency to the miniaturisation of the on-board components point to small, mini and micro UAVs with increasing capabilities. Aerial robotics can be considered in the confluence of different technologies including aeronautics, embedded control systems, communication, and intelligent systems.

However, the flight and payload limitations of affordable UAVs are still significant constraints for the development of many applications. The multiple UAV approach increases the spatial coverage, improves the reliability due to redundancies, allows the simultaneous intervention in different places, and makes possible the teaming of specialised UAVs. Thus, the use of multiple UAVs is a useful approach for many applications.

This book addresses the current state of the technology, existing problems and potentialities of systems consisting on multiple UAVs, emphasizing the teams of heterogeneous UAVs. Heterogeneity here refers to different characteristics of the aerial vehicles, different on-board payloads, and different on-board information processing capabilities ranging from pure remotely teleoperated vehicles to fully autonomous aerial robots.

This book benefits from the results of the COMETS project devoted to the coordination and control of multiple heterogeneous UAVs. This has been a successful project funded by the Information Society Technologies (IST) Research and Development Programme of the European Commission. COMETS has been recognized as the first project that has demonstrated experimentally the cooperation of multiple heterogeneous UAVs for civil applications in Europe. The project generated a new control architecture integrating distributed sensing, real-time image processing and autonomous, fault-tolerant navigation techniques. In COMETS, the cooperation of the vehicles was essential to achieve maximum functionality, geographical diversity and autonomy whilst keeping the operational overhead at a minimum.

However, the book goes beyond COMETS and also considers the research and development activities in multi-UAVs in the framework of the emerging aerial robotics scene.

The book approaches multi-UAV systems from the concepts and theory involved in decisional architectures and cooperative perception down to the UAV prototypes and the detailed consideration of a relevant application: forest fire detection and monitoring. This is a relevant application of multi-UAV systems with implications in the protection of people and the environment.

The book consists of five main parts. The first one is the introductory Chapter that includes an overview of aerial robotics applications. The second part (Chaps. 2–4) is devoted to the methods and technologies in multi-UAV systems. It includes decisional architectures (Chap. 2), communication (Chap. 3) and cooperative perception (Chap. 4) technologies. The third part (Chaps. 5–7) deals with particular UAV prototypes and technologies including autonomous helicopters (Chap. 5), autonomous airships (Chap. 6), and teleoperation tools for UAV guidance and integration in multi-UAV systems (Chap. 7). The fourth part of the book (Chap. 8) is devoted to the forest fire detection and monitoring application case. Finally, the book includes a Chapter with conclusions which also points to future research, developments and applications.

It would be not having been possible to produce this book without the help of different people. We are grateful to all the persons involved in the COMETS project, which are not authors of the book chapters, including Luis Gonzalo Gutiérrez, Francisco Colmenero and Iker Veiga (GMV), as well as Miguel Angel González and Francisca Gómez Pantoja (Helivision), and members of other partner teams in AICIA-University of Seville, LAAS-CNRS, Technical University of Berlin, ADAI-University of Coimbra, and University of Linköping.

We would also like to thank the Officers of the European Commission, and particularly Rolf Riemenschneider for his excellent work in conducting the COMETS project. Furthermore we would like here to recognize the excellent work of the COMETS project reviewers: Fernando Lobo Pereira, Ørnulf Jan Rødseth, and Mario Innocenti for their very constructive advice and orientation of the project.

In addition of COMETS, the work in this book has been supported by the funding of other National projects such as the Spanish AEROSENS and SADCON, and the German DFG grant HO 1257/26-1/2: "Universelles Transportsystem auf der Basis mehrerer autonom fliegender Hubschrauber".

Seville, Spain,
May 2007

Aníbal Ollero
Iván Maza

Contents

1 Introduction
Aníbal Ollero, Iván Maza 1

2 Decision Making in Multi-UAVs Systems: Architecture and Algorithms
Simon Lacroix, Rachid Alami, Thomas Lemaire, Gautier Hattenberger, Jérémi Gancet 15

3 Communication
Volker Remuß, Marek Musial, Günter Hommel 49

4 Multi-UAV Cooperative Perception Techniques
Luis Merino, Fernando Caballero, Joaquín Ferruz, Johan Wiklund, Per-Erik Forssén, Aníbal Ollero 67

5 Autonomous Helicopters
Volker Remuß, Carsten Deeg, Marek Musial, Günter Hommel, Manuel Béjar, Federico Cuesta, Aníbal Ollero 111

6 Airship Control
Leonardo Solaque, Simon Lacroix 147

7 Teleoperation Tools
Nicolás Peña, Federico Cuesta, Aníbal Ollero, Carlos Nogales, Francisco López-Pichaco 189

8 Multi-UAV Experiments: Application to Forest Fires
J.R. Martínez-de-Dios, Luis Merino, Aníbal Ollero, Luis M. Ribeiro, Xavier Viegas 207

9 Conclusions and Future Directions
Aníbal Ollero, Iván Maza 229

Author Index 233

List of Contributors

Aníbal Ollero
Robotics, Vision and Control Group,
University of Seville
Camino de los Descubrimientos s/n
Seville - Spain, 41092
aollero@cartuja.us.es

Iván Maza
Robotics, Vision and Control Group,
University of Seville
Camino de los Descubrimientos s/n
Seville - Spain, 41092
imaza@cartuja.us.es

Simon Lacroix
LAAS-CNRS
7 avenue du colonel Roche
Toulouse - France, 31400
simon.lacroix@laas.fr

Rachid Alami
LAAS-CNRS
7 avenue du colonel Roche
Toulouse - France, 31400
rachid.alami@laas.fr

Thomas Lemaire
LAAS-CNRS
7 avenue du colonel Roche
Toulouse - France, 31400
thomas.lemaire@laas.fr

Gautier Hattenberger
LAAS-CNRS
7 avenue du colonel Roche
Toulouse - France, 31400
gautier.hattenberger@laas.fr

Jérémi Gancet
LAAS-CNRS
7 avenue du colonel Roche
Toulouse - France, 31400
jeremi.gancet@laas.fr

Volker Remuß
Technische Universität Berlin
Strasse des 17. Juni 135
Berlin - Germany, 10623
remuss@cs.tu-berlin.de

Marek Musial
Technische Universität Berlin
Strasse des 17. Juni 135
Berlin - Germany, 10623
musial@cs.tu-berlin.de

Günter Hommel
Technische Universität Berlin
Strasse des 17. Juni 135
Berlin - Germany, 10623
hommel@cs.tu-berlin.de

Luis Merino
Pablo de Olavide University
Crta. Utrera km. 1
Seville - Spain, 41013
lmercab@upo.es

Fernando Caballero
Robotics, Vision and Control Group,
University of Seville
Camino de los Descubrimientos s/n
Seville - Spain, 41092
caba@cartuja.us.es

Joaquín Ferruz
Robotics, Vision and Control Group,
University of Seville
Camino de los Descubrimientos s/n
Seville - Spain, 41092
ferruz@cartuja.us.es

Johan Wiklund
Linköping University
SE-581 83
Linköping - Sweden
jowi@@isy.liu.se

Per-Erik Forssén
Linköping University
SE-581 83
Linköping - Sweden
perfo@@isy.liu.se

Carsten Deeg
Technische Universität Berlin
Strasse des 17. Juni 135
Berlin - Germany, 10623
deeg@cs.tu-berlin.de

Manuel Béjar
Pablo de Olavide University
Crta. Utrera km. 1
Seville - Spain, 41013
mbejdom@upo.es

Federico Cuesta
Robotics, Vision and Control Group,
University of Seville
Camino de los Descubrimientos s/n
Seville - Spain, 41092
fede@cartuja.us.es

Leonardo Solaque
De los Andes University
Carrera 1ra No. 18A-10
Bogotá (Colombia)
l-solaqu@uniandes.edu.co

Nicolás Peña
Robotics, Vision and Control Group,
University of Seville
Camino de los Descubrimientos s/n
Seville - Spain, 41092
nicolasp@cartuja.us.es

Carlos Nogales
Robotics, Vision and Control Group,
University of Seville
Camino de los Descubrimientos s/n
Seville - Spain, 41092
nogales@cartuja.us.es

Francisco López-Pichaco
Robotics, Vision and Control Group,
University of Seville
Camino de los Descubrimientos s/n
Seville - Spain, 41092
pichaco@cartuja.us.es

J.R. Martínez-de-Dios
Robotics, Vision and Control Group,
University of Seville
Camino de los Descubrimientos s/n
Seville - Spain, 41092
jdedios@cartuja.us.es

Luis M. Ribeiro
Universidade de Coimbra
Apartado 10131
Coimbra - Portugal, 3030
luisr1@netcabo.pt

Xavier Viegas
Universidade de Coimbra
Apartado 10131
Coimbra - Portugal, 3030
xavier.viegas@dem.uc.pt

1

Introduction

Aníbal Ollero and Iván Maza

Robotics, Vision and Control Group, University of Seville, Camino de los Descubrimientos s/n, 41092 Seville (Spain)
{aollero,imaza}@cartuja.us.es

Summary. This Chapter deals with the practical interest of Unmanned Aerial Vehicles (UAVs) and aerial robotics. The Chapter also introduces multi-UAV systems and some projects carried out in the last years. Moreover, it also reviews the main civilian applications of UAVs. Finally, the Chapter includes a short overview of the book.

1.1 Aerial Robotics

Outdoor robotics applications in natural terrains usually require higher mobility than the motion capabilities provided by existing ground robotic vehicles. In fact, in spite of the progress in the development of unmanned ground vehicles in the last 20 years, the navigation in unstructured natural terrains still poses significant challenges. Thus, the existing ground vehicles have inherent limitations to reach the desired locations to sense and actuate in many applications. The characteristics of the terrain and the presence of obstacles, together with the requirement of fast response, may impose a major drawback to the use of any ground locomotion system. Then, in many cases, the use of aerial vehicles is the only way to approach the targets to get information or to deploy instrumentation.

Unmanned Aerial Vehicles (UAVs) offer many opportunities to be used in robotic systems avoiding the above mentioned limitations of the ground vehicles. In the last ten years UAVs improved their autonomy both in energy and information processing. Significant achievements have been obtained in autonomous positioning and tracking. These improvements are based on modern satellite based position technologies, inertial navigation systems, communication and control technologies, and image processing. Furthermore, new sensing and processing capabilities have been implemented on-board the UAVs. Thus, today we can consider some UAVs as intelligent robotic systems integrating perception, learning, real-time control, situation assessment, reasoning, decision-making and planning capabilities for evolving and operating in complex environments.

1.2 Multi-UAV Systems

The development of small and relatively low cost UAVs is fuelled by the progress of embedded systems together with the developments on technologies for integration and miniaturization.

A. Ollero and I. Maza (Eds.): Mult. Hetero. Unmanned Aerial Vehi., STAR 37, pp. 1–14, 2007.
springerlink.com

However, constraints such as power consumption, weight and size play an important role in UAVs, and particularly in small size, light and low cost UAVs. Then, the cooperation of many of these vehicles is the most suitable approach for many applications.

A single powerful aerial vehicle equipped with a large array of different sensors of different modalities is limited at any given instant in time to a single view point. However, a team of aerial vehicles can simultaneously collect information from multiple locations and exploit the information derived from multiple disparate points to build models that can be used to take decisions. Furthermore, having a team with multiple heterogeneous aerial vehicles offers additional advantages due to the possibility of exploiting the complementarities of vehicles such as helicopters, airplanes and airships.

The coordination of a team of autonomous aerial vehicles offers many application possibilities. Team members can exchange sensor information, collaborate to track and identify targets and perform detection and monitoring activities among other tasks. Thus, for example, a team of aerial vehicles can be used for detection, precise localization, monitoring and measurement of the evolution of natural disasters, such as forest fires. Furthermore, this approach leads to redundant solutions offering greater fault tolerance and flexibility.

Getting groups of UAVs to team (and small UAVs to swarm) in order to accomplish a given objective requires significant coordination efforts. Several existing methods are related to the control of homogeneous teams of airplanes for military applications involving problems such as the close-formation flight (see for example [5, 6, 9]) and the autonomous aerobatic manoeuvring (Phoenix Project at Princeton University). The tendency is towards the development of systems that do not require huge computational overhead or large communication bandwidth. However, most studies have been conducted only in simulation, and the demonstrations with real aircrafts are still very scarce.

At MIT, a Multi-vehicle Experimental Platform for Distributed Coordination and Control is being developed [1]. This project will build a multi-vehicle testbed to demonstrate and evaluate the coordination and control approaches under development at MIT as part of several ongoing DoD funded research programs. A demonstration with two autonomous UAVs flying in formation has been performed.

The ANSER project of the Australian Centre for Field Robotics is aimed at demonstrating Decentralised Data Fusion (DDF) and Simultaneous Localisation and Mapping (SLAM) with multiple cooperative UAVs. The demonstration of DDF and SLAM requires both navigation and terrain sensors to be carried onboard the UAVs. These include an INS/GPS navigation system, millimeter wave (MMW) radar, and both a single vision node and a vision node augmented with a laser system.

The goal of the STTR program, jointly developed by the Cornell University and the Insitu Group, is to evaluate and demonstrate algorithms for cooperative tracking of moving targets using multiple UAVs. This program develops planning, control, and cooperative tracking estimation algorithms for the ScanEagle and SeaScan UAVs.

Most of the research work has been devoted to homogeneous teams of aerial vehicles. The cooperation of heterogeneous aerial vehicles offers significant benefits because of the exploitation of complementarities among vehicles with different mobility attributes and also different sensors (with different associated perception functionalities), but also poses significant challenges due to the complexity of the involved problems.

This book is devoted to consider the current state of the technology, existing problems and potentialities of a system consisting on multiple unmanned aerial vehicles, emphasizing systems composed of heterogeneous UAVs. This heterogeneity is two-fold. Firstly, it considers the heterogeneity in the UAV platforms, which is usually aimed at exploiting the different characteristics of the different aerial vehicles (such as helicopters and airships). Secondly, the heterogeneity present in the information processing capabilities on-board, which could range from pure remotely teleoperated vehicles to fully autonomous aerial robots, is considered.

1.3 Applications

Currently the main UAV applications are defense related and the main investments are driven by future military scenarios. Today the civilian UAV market is small compared to the military one. Significant civil markets for UAVs are still to emerge, with only limited niche applications being currently available. However, in the next 10–15 years, the expectations for market growth of civil and commercial aerial robotics are very high. In the following, some of these applications are reviewed.

1.3.1 Aerial Photography and Cinematography

Many UAV applications are related to the acquisition of aerial images. Thus, aerial photography and cinematography are today applications with growing markets. Nowadays, conventional Remotely Piloted Vehicles (RPVs) are being used. Small fixed wing and airship UAVs can be used for aerial photography. Cinematography requires the application of helicopters because their greater maneuvering capabilities. However, when using conventional RPVs, the distance to the pilot should be very short and direct view of the aerial vehicle is required. The autonomous aerial vehicles do not have these constraints. Advanced teleoperation systems can be applied to improve the teleoperation performance. Thus, on-board cameras can provide to the human remote pilot the vision for the operation out of the line of sight. These images can be displayed using glasses and other headed mounted displays, or portable displays in general. Moreover, if the map is available, the ground teleoperation station can display the UAV in the map, eventually with 3D representations of both the environment map and the UAV, facilitating the remote piloting task. This "out of the line-of-sight" remote piloting is suitable for airplanes. However, it is not enough for helicopters, because of the required high frequency stabilization loops closed by the human pilot of remotely piloted helicopters. Then, on-board control systems for autonomous helicopter stabilization have been implemented. The above mentioned teleoperation strategies could be enough for some applications or could be understood as a first step towards autonomy.

Today, the most suitable strategy for aerial photography is the combination of low level local autonomy in the on-board controller and a guidance by the pilot that could be eventually implemented using conventional joystick-type hand controllers.

1.3.2 Aerial Mapping

Aerial mapping is another application with potential short term economical impact. The autonomous aerial systems can build more accurate terrain maps than conventional aircrafts at substantial cost savings. Unlike the conventional aerial piloted vehicles, the helicopters and airships can fly close to the ground while carrying cameras or range sensors to build high-resolution 3D maps. Furthermore, the helicopters can fly in smaller and more constrained areas to build highly detailed maps.

Terrain mapping was also one of the goal missions of the COMETS project. It consists in providing an update of the initially available cartographic data bases of the area over which the UAVs are operating. This process provides up-to-date information at a higher resolution than the initially available data bases.

Terrain mapping can be useful for many other applications considered in this section such as environmental monitoring, agriculture and forestry (inventories), inspection, surveillance, disaster and crisis management (asses areas affected by disasters).

The maps can be also needed for civil work or even as a support for the UAV operation. As opposed to a live video stream, a terrain map provides a spatially coherent view of the scene that memorizes the areas over which the UAVs flew.

Figure 1.1 illustrates the generation of a high resolution map using an autonomous blimp.

1.3.3 Meteorology

Some UAVs have been already used for meteorological missions. Recently, solar-powered UAVs for high-altitude, long-endurance flights have been developed (HALEs). Within a few years, solar-powered UAVs, incorporating energy storage for night-time flight, will be capable of continuous flight for months at a time at altitudes of over 60,000 feet. Applications for such UAVs include telecommunications, reconnaissance and atmospheric measurement.

1.3.4 Environmental Monitoring

Environmental monitoring is another potential application for UAVs with imaging and other sensors for the measurement of environmental variables. This, includes oil spill discovery and monitoring.

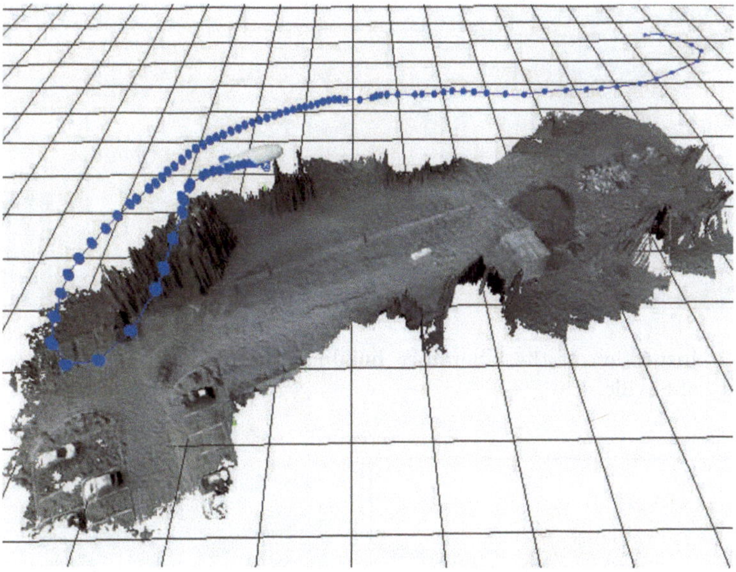

Fig. 1.1. Building a digital terrain map using an autonomous blimp (courtesy of LAAS-CNRS)

These applications require HALE UAVs or medium-altitude, long-endurance (MALE) UAVs with significant autonomy. The advantages when comparing with today existing satellite based technologies are the higher spatial and temporal resolution than can be obtained with UAVs. Furthermore, it is also possible to apply small UAVs with lower endurance to collect very high resolution information in local sites. Thus, for example, in [2] the architecture of a small fixed wing UAV for volcanic gas sampling is presented.

1.3.5 Agriculture and Forestry

UAVs can also be used in agriculture and forestry. This includes generating an inventory of agricultural areas or forestry, as well as chemical spraying and dusting in large agricultural areas. Thus, crop chemical spraying is a well known application currently performed by autonomous helicopters in Japan.

1.3.6 Inspection

Another potential application for small UAVs, with significant short and medium term possibilities, is the inspection of buildings and utilities. The recently concluded Helinspec project demonstrated the thermal inspection of buildings by using a helicopter UAVs with infrared and visual cameras [3] (see Fig. 1.2). The perception system is based on the use of an infrared camera and a visual video camera. Infrared images are processed for the detection of building anomalies,

Fig. 1.2. Inspection of the laboratory buildings of the School of Engineers of the University of Seville

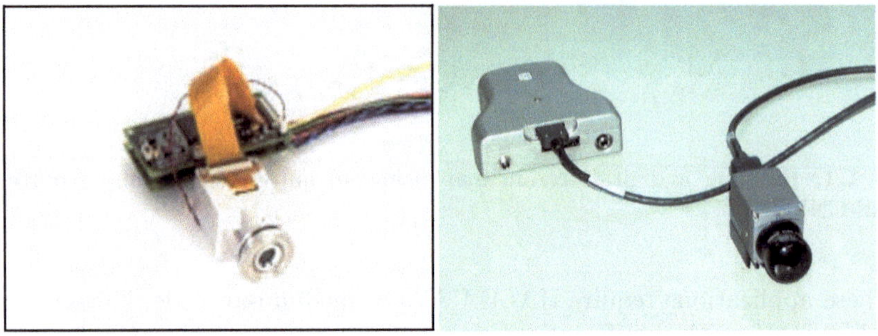

Fig. 1.3. Raytheon 2000AS and Indigo Omega micro cameras used for infrared building inspection

whereas visual images are used for better interpreting the anomalies. To facilitate their joint processing, infrared and visual images are synchronized using time stamps. Two micro infrared cameras were used: a Raytheon 2000AS and an Indigo Omega (see Fig. 1.3). Both cameras weight approximately 120 grams and operate in the far-infrared band (7–14 microns).

The Indigo Omega is a thermographic infrared camera and provides images via the IEEE 1394 protocol whereas the Raytheon 2000AS provide non-thermographic CCIR images. The visual camera is a low-weight color device. Figure 1.4 shows four non-stabilised infrared images taken from the UAV in hovering. Figure 1.5 shows the results after applying image stabilisation and thermal anomaly detection by using the methods presented in [3]. The Regions Of Interest (ROI) detected are marked with a white rectangle overlaid on the images. These results were validated by manual inspection with a ground infrared camera (FLIR THERMACAM P20).

There are also projects on the inspection of electrical power lines, pipelines and other utilities. These and other inspection tasks currently performed by traditional aircrafts could be affected in the future.

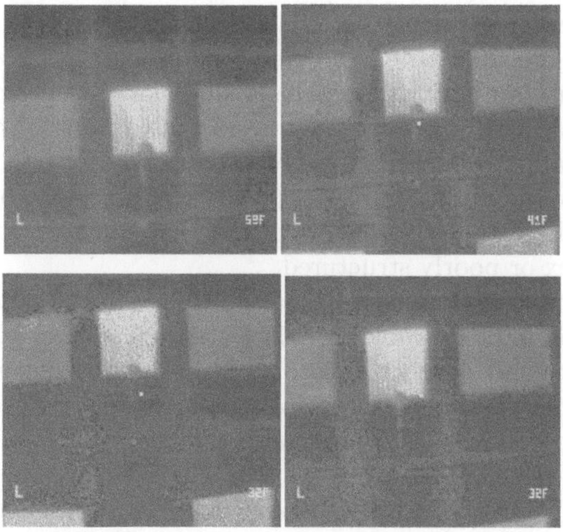

Fig. 1.4. Four non-stabilised infrared images taken with an Indigo Omega from a UAV in hovering

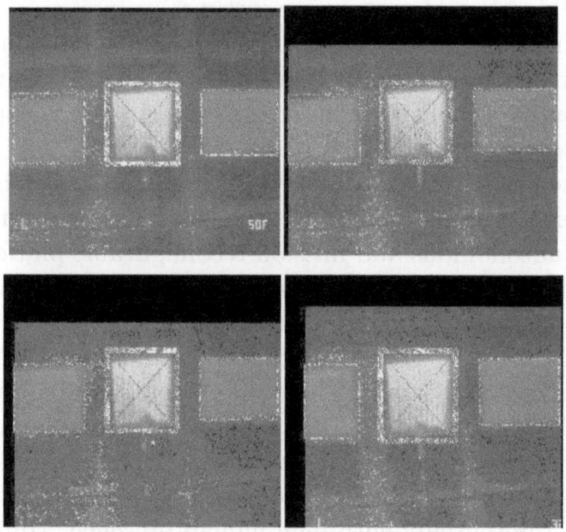

Fig. 1.5. Images after image stabilisation and thermal anomaly detection

1.3.7 Law Enforcement and Security Applications

In the coming years, security enforcement will rely more and more on automated systems. Thus, domestic (homeland) security is one of the main civil UAV application drivers. Medium term applications could include surveillance of critical infrastructures (telecommunication and power lines, water, gas and oil pipelines, etc.) and extended border and coastal patrol boosted by the need to monitor illegal immigration fluxes.

Law enforcement in general is an application where the UAVs can collaborate to aid the police in criminal search operations providing images crucial to the tactical assessment of the situation. This application includes crime monitoring (crowd control, car and boat surveillance and chase operations, etc.) as well as border and coastal patrolling.

Autonomous systems in security applications will need to operate in hostile, tedious, or hard-to-access environments that are partially or completely unknown, complex or poorly structured.

Complex security missions will increasingly require the deployment and cooperation of many inter-operable robotic systems that inter-communicate, and closely collaborate and interact. Thus, the coordination and even cooperation of many autonomous systems will play a significant role in these missions. This includes the cooperation of UAVs but also the cooperation of UAVs with Unmanned Ground Vehicles (UGVs).

As mentioned above, border surveillance is a security scenario where the application of UAVs provides significant advantages. There are several initiatives on the short term implementation of this application, i.e. in the eastern borders of Europe. This is also a critical application along the south coast of Spain demanding the development of new surveillance technologies that could be applied not only for searching but also to rescue victims.

Border patrolling requires a high level of autonomy in mission management together with advanced cognitive capabilities. Furthermore, the natural tendency is towards distributed applications involving different range capacity UAVs and UGVs, fixed infrastructures and command station operators. In these applications the robotic teams will operate and evolve in dynamically changing, partially unknown, complex and sometime hostile or hazardous environment and this will require modular, autonomous and mission-reconfigurable platforms.

Another application is site protection ranging from industrial sites to public sites. There are some common characteristics with the border surveillance, but here UAVs operate on a more local level. Furthermore, they have to cooperate with indoors autonomous systems. These robots will be integrated in the overall surveillance team, which includes security patrols and a command station, to fulfill tasks such as: surveillance, threat and illegal activities detection, dissuasive action, entrance control, etc. The environment of operation is one of high dynamic changes and potentially with high human presence. Mini to micro UAVs, UGVs and "human-friendly" robots are envisioned for such a scenario.

Other longer term application that can be included in security enforcement is maritime surveillance, involving maritime traffic control and monitoring of ship movements, supervising (illegal) fishery, search and rescue operations, etc, and surveillance of hazardous material for tracking and escorting highly sensitive individual shipments to their destinations.

1.3.8 Disasters and Crisis Management

UAVs could be also useful to protect people in case of natural disasters. Thus, for example, the UAVs can collaborate to quickly and systematically explore the affected area to locate victims. They can then visually lock on to objects or victims at the site to guide rescue forces. Furthermore, in some cases, they can be more readily deployed in weather conditions that would normally prevent human piloted search, and they can be sacrificed in very dangerous conditions to save human lives.

Some applications in crisis management are related to the above mentioned possible commercial applications. Thus, UAVs can be also very useful to inspect buildings and roads for potential damage after an earthquake, and to build detailed terrain maps of the affected areas. Here it should be mentioned the application of UAVs after the Kathrina disaster in USA [7].

1.3.9 Fire Fighting

This application could be included in the "Natural Disasters and Crisis Management". However, it also has some characteristics in common with "Environmental Monitoring".

Forest fire fighting including prevention activities before the fire, detection, monitoring and post-fire analysis has been mentioned as one of the suitable short time applications of UAVs.

UAVs can be applied for forest surveillance improving the spatial and temporal resolution of satellite observations in activities such as building of fire risk maps, monitoring of the vegetation and even generating 3D vegetation maps with scanning sensors such as LIDARs. Some current HALE UAVs and future stratospheric UAVs are suitable for these tasks due to their high endurance and high payload capable of carrying relatively heavy observation sensors. Moreover, small and low-cost UAVs can be also used for local high resolution mapping, which can be useful in special protected areas and also to support forest fire research and development activities.

UAVs can also be used for automatic forest fire detection improving the coverage of ground automatic systems and the temporal and spatial resolution of satellite based systems. Stratospheric, HALE and MALE UAVs can provide data with higher frequency and resolution. Furthermore, tactical UAVs and small low cost UAVs can be used to confirm and localize the alarms, as well as to monitor the fire, when launched nearby. Figure 1.6 shows an aerial photograph taken during a controlled forest fire in Portugal. Using this kind of images, the fire front can be precisely localized and a computer system can also provide fire parameters such as the spread velocity and the height of the flames.

UAVs can be used also for post-fire analysis. MALE and tactical UAVs can be applied to detect the presence of active fire embers. Other tasks carried out during this stage are burnt-area mapping.

The cooperation of several UAVs has also many potentialities in forest-fire activities [8]. For instance, different UAVs of a fleet can survey different regions at the same time, allowing the establishment of priorities in the surveillance according to the interest or risk of the region. Different types of UAVs can play a different role. Thus, for example, tactical UAVs or mini-UAVs can be used to confirm and precisely localize alarms detected by the High-Altitude and Long

Fig. 1.6. Aerial photograph taken during a controlled forest fire in Portugal (courtesy of ADAI)

Endurance (HALE) UAVs. Moreover, the complementary of data gathered by different UAVs can be used for confirmation of the fire alarms and thus can increase the robustness of the system.

The COMETS project has demonstrated the interest of multi-UAV systems in forest fire activities.

1.3.10 Traffic Surveillance

This application has been analyzed in the WITAS project of the Linköping University (Sweden) and also in the COMETS project. The intended operational environment is over an area with widely varying geographical terrain containing traffic networks in city, suburban and rural areas, which may be both densely and sparsely populated by vehicles ranging from passenger cars to tractor-trailers (see Fig. 1.7).

The UAV is required to "understand" what happens on those roads – conventional maneuvers of individual cars and other road vehicles, dangerous or otherwise exceptional maneuvers, structure of the traffic (e.g. congestion), etc. It is also required to perform tasks that are assigned by the operator or triggered by the observations it makes itself, for example to follow a certain car that flees from the scene of an apparent crime, or to assist a certain car so that it can make it through difficult traffic and get to a particular destination as quickly as possible.

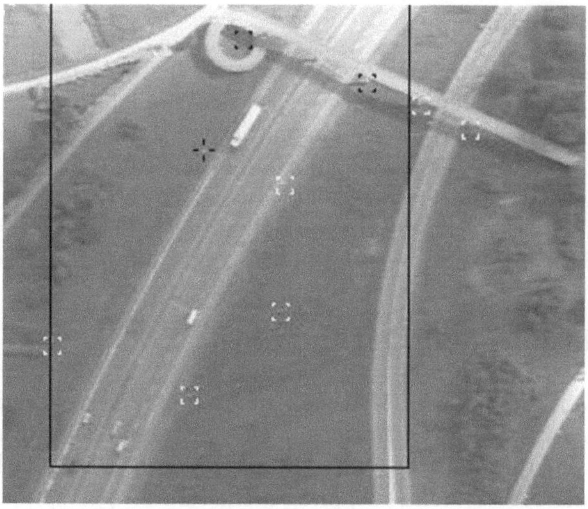

Fig. 1.7. Traffic surveillance using an UAV (courtesy of Linköping University)

The activities in this operational environment will be those that one would expect from familiarity with normal road traffic. The UAV's "understanding" of the environment is dictated by its sensory capabilities and by the major tasks that are defined for it. They are: to monitor traffic situations, to identify and track individual vehicles, to identify episodic behavior of both individual and groups of vehicles, to gather data pertaining to road network use and abuse, to provide assistance to emergency services, and to serve as a mobile sensory platform with real-time information gathering and processing capabilities.

One important basic processing step is to find moving objects in the scene. This can be a challenging task due to camera motion etc. For example, motion blobs can be detected via spatio-temporal filtering. These motion blobs contain information of motion velocity and direction, and can be used as input to an object tracking algorithm. From the motion information of the individual vehicles, it is possible to implement the tasks described above.

1.3.11 Communications

Other commercial applications of UAVs (HALE and MALE) is communication serving as broadcasting platforms or as cellular relays, particularly in areas without communication infrastructure or where this infrastructure is damaged or not available for other reasons.

Thus, an UAV could be placed hovering or near-hovering at high altitude acting as a repeater for different communicating systems [4]. Several projects around the world deal with HALE UAVs for communications, as the SkyNet initiative in Japan, the finished HeliNet project [10] funded by the European Commission, the Solitair platform by DLR in Germany and others.

1.3.12 Civil Engineering

Civil works usually require aerial views to asses the state of the activities and eventually determine deviations with respect to the planned work.

These images are also required in case of the construction of industrial complexes or even urban buildings and can be obtained from small UAVs with substantial saving costs. Furthermore, helicopter UAVs can also easily provide lateral views without the risk of flying near buildings. The high resolution maps mentioned before are also very useful in this application.

1.4 Conclusions and Outline of the Book

Aerial robotics is an emergent field with many potential applications. In many of these applications the cooperation of multiple UAVs is very relevant. This book presents different UAV methods and technologies. Particularly, the book is devoted to multi-UAV systems including architectures, perception, communication, and control methods. Finally, Chapter 8 presents multi-UAV experiments and the application to forest fires.

This book benefits from the results of the recently concluded COMETS project. This has been a successful project funded by the IST programme of the European Commission devoted to the coordination and control of multiple heterogeneous UAVs. However, the book will go beyond COMETS and will also review the research and development activities in multi-UAVs in the framework of the emerging *aerial robotics* scene.

Figure 1.8 shows a general schema which clarifies the organization and structure of the book. Three main blocks can be identified (multi-UAV methods and technologies, single-UAV methods and technologies and applications) and the chapters corresponding to each block are shown.

The first chapter of the book has provided a general introduction to multiple autonomous aerial vehicles by including their application potentialities.

The second chapter is devoted to decision making. Different multi-robot planning methods are reviewed. Then, an architecture for the cooperation of heterogeneous UAVs is presented, and results on multi-UAV planning are depicted.

Chapter 3 addresses the particularities of the communication system required for the cooperation. Moreover, a software communication system, which can be applied by using different hardware implementations, is described.

Multi-UAV perception techniques is the subject of Chap. 4. The chapter addresses general cooperative perception issues involving detection and monitoring applications.

The next part of the book includes three chapters addressing UAV methods and technologies. Thus, Chaps. 5 and 6 are devoted to autonomous helicopters and blimps respectively, including modelling and control. Then, Chap. 7 is devoted to UAV teleoperation tools. Particularly, a system for the teleoperation of helicopters and its integration in a multi-UAV system is presented.

The next chapter of the book is devoted to a case study that will illustrate the application potentialities of the multi-UAV system. The selected application is

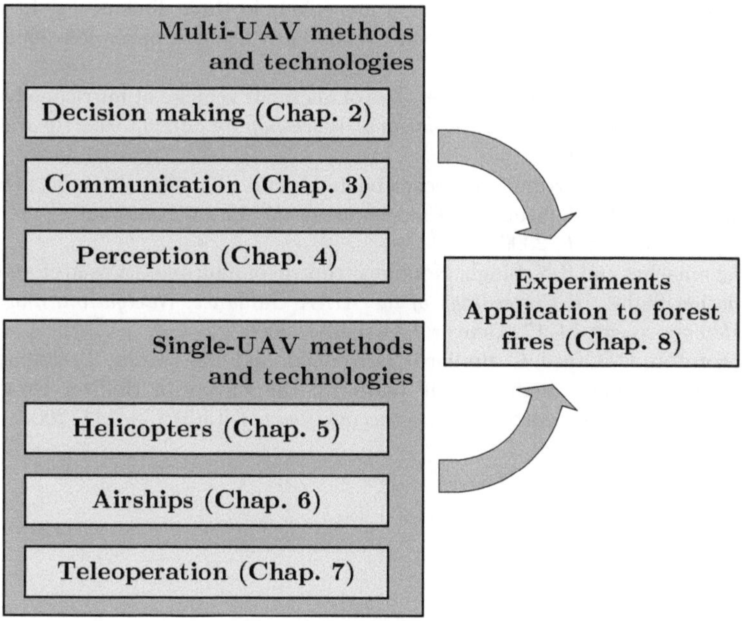

Fig. 1.8. Organization of the book

forest fire detection and monitoring. The chapter will present the results obtained in the experiments carried out with controlled fires in the framework of the COMETS project.

Finally, Chap. 9 will present conclusions and will point out future research, development and applications.

References

1. Multi-vehicle experimental platform for distributed coordination and control (DoD DURIP). `http://www.mit.edu/people/jhow/durip1.html`.
2. D. Caltabiano, G.Muscato, A.Orlando, C. Federico, G. Giudice, and S. Guerrieri. Architecture of a UAV for volcanic gas sampling. In *Proceedings of the 10th IEEE International Conference on Emerging Technologies and Factory Automation*, Catania, Italy, September 2005.
3. J.R. Martinez de Dios, A. Ollero, and J. Ferruz. Infrared inspection of building using autonomous helicopters. In *Proceedings of the IFAC Symposium on Mechatronics*, Heidelberg, Germany, 2006.
4. G. M. Djuknic, L. J. Freidenfelds, and Y. Okunev. Establishing wireless communications services via high-altitude platforms: a concept whose time has come? *IEEE Communications Magazine*, 35(9):128–135, September 1997.
5. F. Giulietti, L. Pollini, and M. Innocenti. Autonomous formation flight. *IEEE Control Systems Magazine*, 20(6):34–44, December 2000.

6. J.K. Hall and M. Pachter. Formation maneuvers in three dimensions. In *Proceedings of the 39th IEEE Conference on Decision and Control*, pages 364–369, Sydney, Australia, December 2000.
7. R. Murphy, J. Burke, and S. Stover. Use of micro air vehicles at hurricane Kathrina. In *Proceedings of the IEEE International Workshop on Safety, Security and Rescue Robotics*, August 2006.
8. A. Ollero, J. R. Martinez de Dios, and L Merino. Unmanned aerial vehicles as tools for forest-fire fighting. In *Proceedings of the VI International Conference on Forest Fire Research*, 2006.
9. C. Schumacher and S.N. Singh. Nonlinear control of multiple UAVs in close-coupled formation flight. In *Proceedings of the AIAA Guidance, Navigation, and Control Conference*, pages 14–17, Denver, CO (USA), 2000.
10. J. Thornton, D. Grace, C. Spillard, T. Konefal, and T.C. Tozer. Broadband communications from a high-altitude platform: the European HeliNet Programme. *Electronics & Communication Engineering Journal*, 13(3):138–144, 2001.

2

Decision Making in Multi-UAVs Systems: Architecture and Algorithms

Simon Lacroix, Rachid Alami, Thomas Lemaire,
Gautier Hattenberger, and Jérémi Gancet

LAAS-CNRS, 7 avenue du colonel Roche, 31400 Toulouse (France)
firstname.name@laas.fr

Summary. This chapter depicts an architecture that aims at designing a multi-UAV framework enabling cooperative operations in a system in which some UAVs are directly controlled by an operator, others are only endowed with *operational autonomy*, and others have *decisional autonomy* capacities. The architecture provides with the possibility to configure dynamically the decisional scheme, depending on the available robots and on the operational context.

A taxonomy of robots decisional autonomy is introduced, and used as a foundation to state the proposed architecture. The various functionalities on-board each robot are organized among a repartition that exhibits on-board functional components, and on-board or on-ground generic executive and decision making processes.

A set of algorithms that fulfill the three main decision-making functionalities required in a multi-robot system are then presented: a contract-net protocol that can handle task allocation for complex multi-UAV missions, a planning scheme based on a Hierarchical Task Networks planner completed with plan-refiners that consider the actual domain models, and an executive system that handles the coordination and task execution issues.

2.1 Introduction

Several UAV projects have been led by different research teams. Many are focused mainly on the development of advanced flight control abilities, e.g. to achieve aggressive maneuvers with helicopters, and others rely on operational UAV autonomy, i.e. the UAVs receive a pre-planned sequence of tasks to achieve, and do not exhibit high level planning or decisional skills – a noticeable exception being the WITAS project at Linköping University [17]. When it comes to multi-UAVs, various topics have been studied. In the ANSER project [38], the focus is set on the data fusion between multiple vehicles, e.g. to recover the positions of features on the ground [37].

Several contributions tackle the formation flight problem according to a control theoretic approach [10, 32], or using reactive behavior-based controllers in [6]. For this problem, the deliberative activities consist in determining the trajectories to be followed by the formation (according to either a centralized [24] or distributed

A. Ollero and I. Maza (Eds.): Mult. Hetero. Unmanned Aerial Vehi., STAR 37, pp. 15–48, 2007.
springerlink.com

paradigm [30]), in selecting the internal formation geometric configuration [21], and in achieving switches between two given geometric configurations [41].

Fewer contributions deal with multi-UAVs problems according to a deliberative paradigm, where the UAVs exhibit cooperative and coordinated activities produced by high level planners, while maintaining reactive abilities during the execution of the plans (e.g. [35, 39, 23]).

Problem Statement

We are interested here in the deployment of a multi-UAVs system to achieve observation missions. These missions can consist in detecting particular events over a given area, in monitoring their evolution, or in surveying (mapping) a given area. The system is *controlled by the operators*, in the sense that it achieves missions decided by operators according to their needs and the current knowledge of the situation they have access to. In particular, depending on the situation, a human operator should be able to handle the control of any UAV, at *any control level*. This implies that the operators should be able to specify high level missions, elementary tasks (such as setting a waypoint to reach by a particular UAV), or could even directly control the motions of an UAV. This is an essential feature required in any multi-UAV application context, where the operators need to master the activities of the whole system. Of course, this does not preclude the autonomous achievements of elementary tasks or high level missions: tedious and repetitive operations such as surveying an area relieve the operators if they are performed autonomously. There are even cases where only an advanced planning and control system is able to efficiently succeed, e.g. the coordinated operations of two or more UAVs flying out the operators line of sight is extremely difficult to control remotely.

The main consequence of this "controlability" requirement is that the system must be able to integrate UAVs with various levels of autonomy, from simple teleoperated UAVs to UAVs endowed with mission planning abilities. In other words, the system must enable both centralized (i.e. human-centered, via a central ground station) and distributed (i.e. delegated to UAVs) configurations of the decision. This significantly influences the design and implementation of the decisional architecture and algorithms introduced here.

Approach and Outline

The overall system architecture follows a classic ground station / flying segment scheme. The central ground station is endowed with all the necessary monitoring facilities (in particular, it can include a data processing module that helps the operators to assess the situations). It also has some UAV control means, and mission design and planning abilities.

The ability to integrate *heterogeneous UAVs*[1] and to exhibit adjustable autonomy is mainly brought by the design of the UAV on-board architecture and

[1] The word "heterogeneous" being essentially related to decisional autonomy capabilities.

associated tools. Section 2.2 depicts this architecture, that provides with the possibility to configure dynamically the overall decisional scheme, depending on the available robots and on the operational context. A taxonomy of robots decisional autonomy is introduced, and used as a foundation to state the proposed architecture. It incrementally describes increasing schemes of autonomous decision-making, ranging from no decision-making capabilities up to autonomous task planning capabilities, autonomous coordination and even autonomous dynamic task re-allocation among the UAVs.

Many single UAV missions can be achieved only with operational autonomy abilities: the UAV receives a pre-planned sequence of tasks to achieve, and do not need high level decisional skills. But multi-UAV systems involve more temporal constraints and higher uncertainties on tasks execution: they can require higher autonomy abilities, ranging from coordinated execution control to task allocation. Sections 2.3 to 2.5 presents a set of algorithms that enables various levels of autonomous abilities:

- section 2.3 depicts an executive system that handles the coordination and task execution issues,
- section 2.4 presents a mission planning scheme based on a Hierarchical Task Networks planner completed with plan-refiners that consider the actual domain models,
- and section 2.5 presents a distributed allocation scheme based on a contract-net protocol, that can handle complex multi-UAV missions with temporal constraints between tasks.

2.2 UAV Architecture

Numerous robots architectures have been proposed in the literature. Subsumption architectures [8, 4], exhibit behavioral robots models built upon reactive capabilities, whereas layered models explicitly divide up robots capabilities, as in the three-layers architectures (deliberative, executive and functional) [2, 19], or the 2-tiered CLARAty architecture [40]. Multi-robot architectures embrace additional concerns: designing a multi-robot architecture requires the definition of the decision making scheme and the specification of the interaction framework between the different robots, which of course influences the definition of the individual robots architecture. ALLIANCE [29] provides a behavior-oriented solution, enabling the design of totally distributed, fault tolerant multi-robot systems, whereas Simmons & al. [33] extend the three-layers architecture model within a multi-robot framework, where interactions between robots may occur along the different layers.

Each of these multi-robot architectures enables the coordination and cooperation of several robots, but assumes a given homogeneous level of decisional autonomy for all the robots. They can enable the integration of physically heterogeneous robots, but can not cope with heterogeneous robots in terms of *decisional capabilities*.

2.2.1 A Taxonomy of Decisional Autonomy Capabilities

In any multi-robot system, decisional autonomy encompasses the following four abilities:

- Task allocation: How to distribute tasks among the robots ? This requires to define both a task assignment protocol in the system, and some metrics to assess the relevance of assigning given tasks to such or such robot.
- Mission refinement, planning and scheduling: How to transform a task or a mission into an executable sequence of actions ? These decisional activities are dedicated to plan building, considering models of the actions of the involved UAV, and of the environment.
- Coordination: How to ensure the consistency of the activities within a group of robots ? This requires the definition of mechanisms dedicated to prevent or to solve possible resource conflicts (time and space resources), as well as mechanisms to plan and control the execution of joint cooperative tasks.
- Supervision and execution control: How to ensure that the planned tasks are correctly executed ? A system that manages the task execution and considers the encountered contingencies is required for that purpose.

These decisional components can be implemented according to different multi-robot systems configurations: they can be gathered within a central decisional node, or be partially (or even totally) distributed among the robots themselves. We define the "level of autonomy" of a robot as the amount of decisional components it is endowed with, and consider the following five levels (see Fig. 2.1):

- **Level 1:** no autonomy on-board the robot. The robot is only able to directly execute elementary tasks requested by the central decisional node.
- **Level 2:** executive capabilities. The robot is able to manage partially ordered sequences of elementary tasks, and to return execution status of the tasks.
- **Level 3:** same as level 2, plus coordination capabilities. The robot may manage on-line simple interactions (synchronizations) directly with other robots endowed with at least the same level of decisional autonomy.
- **Level 4:** distributed deliberative capabilities. High level tasks requests are managed (task planning and scheduling), and the multi-robot tasks coordination is autonomously ensured in a distributed way among robots endowed with at least the same level of autonomy.
- **Level 5:** same as level 4, plus tasks re-allocation capabilities. The robots may opportunistically re-allocate tasks and accept new tasks from other robots of the system endowed with the same level of autonomy.

This taxonomy is characterized by a large gap between levels 3 and 4: up to level 3, a Centralized Decisional Node (CDN) is expected to ensure the global consistence of the system's activity: levels 1 to 3 are considered as **"low levels"** of decisional autonomy. Whereas levels 4 and 5 introduce the possibility to delegate coordination and mission refinement activities in a distributed way (**"high levels"** of decisional autonomy, embedded in the Distributed Decision Nodes – DDN),

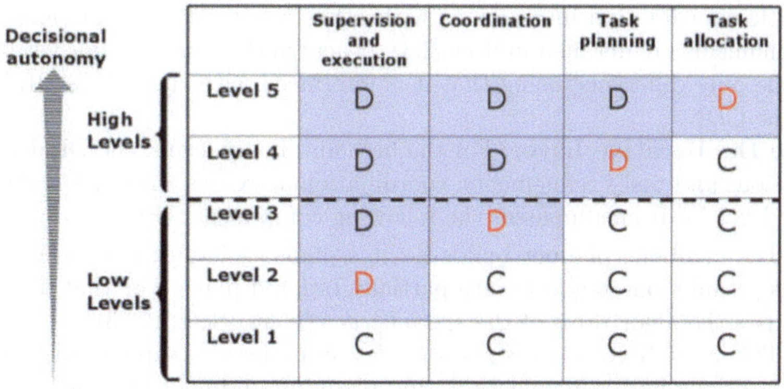

Fig. 2.1. Five levels of decisional autonomy. **C** stands for "Centralized", and **D** stands for "Distributed".

robots belonging to the fifth level being able to dynamically refine the allocation of tasks between them.

This taxonomy is to be understood in terms of *incremental delegation of decisional capabilities* by the multi-UAV system's user towards the UAVs. From the user's point of view, level 1 means a centralized full control of the system (*centralized* should be considered as available for operator). Level 2 enables an autonomous execution of partially ordered plans. Level 3 provides autonomous inter-UAV synchronization capabilities. Then a large gap appears between levels 3 and 4: up to the level 3, the CDN performs tasks planning and ensures the global consistency of the UAVs activities. Whereas level 4 is related to delegating mission refinement and planning activities to the UAV. Finally, level 5 enables autonomous tasks re-allocation : this is the highest delegation of decision making (i.e. the CDN only expresses high level goals to be achieved).

2.2.2 Decisional Architecture

Figure 2.2 depicts the overall architecture of the UAVs, for both low levels and high levels of decisional autonomy. A CDN communicates with robots, exchanging messages whose abstraction is defined according to the robots levels of autonomy. Each robot has a number of functional components, and is endowed with a generic Distributed Decisional Node (DDN) that enables various configurations of decisional autonomy, ranging from the simplest up to the highest decisional capabilities. It encompasses an executive (this executive being actually common to all levels is denoted as the *Multi-Level Executive* - MLE), and a *Deliberative Layer* (DL) which provides robots with higher levels of decisional capabilities.

- **The Multi Level Executive.** For the low levels, the DDN is restricted to an executive. For level 1, the MLE behaves as a transparent connecting point between the CDN and the robot's functional components. For levels 2 and 3, it manages tasks sequences execution, and at level 3 it enables simple coordination interactions with other robots of the same level (these mechanisms are detailed in Sect. 2.3). It acts in the same way for levels 4 and 5, the only difference being that it is interfaced with the UAV's DL instead of the CDN.

- **The Deliberative Layer.** For the high autonomy levels, the DL deals with missions and tasks refinements, coordination activities, and task reallocation (for level 5). It encompasses the following components (Fig. 2.2):
  - The symbolic planner builds flexible plans skeletons: it transforms high level missions requests into partially ordered plans. For that purpose, it uses the algorithms of the *specialized refiners* (Sect. 2.4.3).
  - The specialized refiners gather a set of features to support tasks decompositions and refinements during planning and coordination, relying on the UAVs and the environment's models.
  - The interaction manager provides the means to coordinate UAVs activities, relying on distributed negotiation mechanisms, as the Contract-Net protocol to handle task allocations for instance.
  - The supervisor has a central position in the DL: it transmits missions refinement requests to the symbolic planner, and then triggers negotiation sessions with the interaction manager in order to coordinate the resulting plans. It finally sends plans to be executed towards the MLE, and monitors returned tasks / plans execution status.

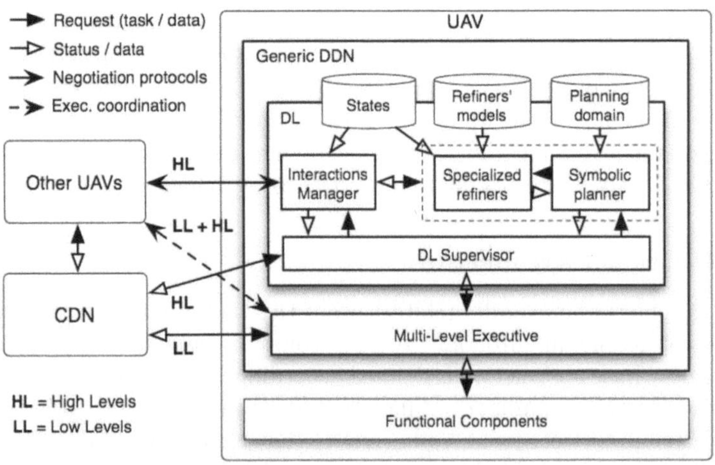

Fig. 2.2. DDN's components

2.3 Execution Control

We focus here on the functionalities of the Multi-Level Executive, using the context of an environment surveillance mission as a support example.

2.3.1 General Task Model and Assumptions

The task model is built around elementary events processing: these events are expected to occur whenever the states of tasks evolves. Events can also correspond to other noticeable activities evolution, such as the reception of a message, or the elapsing of a certain amount of time. Tasks have a temporal extent: a task starts, then ends after a certain amount of time. The starting event is the only controllable event: all other kind of events related to a task are contingent, i.e. the system can not guarantee that such an event will occur, neither exactly when it may occur. A task can give rise to several, partially ordered contingent events during its execution.

For the low levels of autonomy, the Central Decisional Node (CDN) is supposed to be able to elaborate a safe and consistent multi-robot plan, and therefore to provide the robots with the tasks to be processed, according to a task communication formalism. On the other side, the minimal requirement expected from a robot is its capability to execute *elementary tasks*, i.e. unitary "simple" tasks that can be handled by robot's functional components. In the an environment surveillance mission, the following tasks are expected to be processable by a robot integrated in the system: take-off (TO), go-to (GT), take-shot (TS), wait (WT), and land (LD).

Integrating an UAV in the whole system requires to have this UAV endowed with a basic interface enabling elementary tasks information transmission (request, status, execution's result). For that purpose, an *elementary task formalism* has been developed (see Fig. 2.3 – its specification is not necessary to be detailed here).

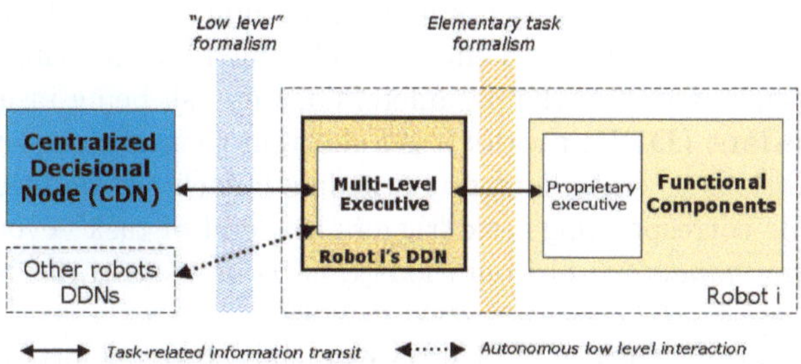

Fig. 2.3. Communication formalisms in the low decisional autonomy levels configurations

2.3.2 Executive's Mechanisms

For the first level of decisional autonomy, the MLE is passive: it only transmits the elementary tasks requested by the CDN to the functional components of the robot, and sends back execution status.

For the second decisional autonomy level, the MLE manages partially ordered sequences of tasks in a consistent way and in a timely and safe manner. Two main mechanisms are involved for this purpose:

- **dynamic tasks insertion:** This enables the possibility to request tasks insertion in the current task plan, according to an insertion mode that will characterize the relative order of the newly inserted task versus the current partial order of the already scheduled tasks. Four possible insertion modes are defined:
  - **SEQuential (SEQ) mode:** This is the most common possibility to insert a new task in the plan. The task has to be provided with a certain number of preconditions (in terms of expected events), which satisfaction can be specified either as *mandatory* or *optional*: in the first case, the satisfiability itself should be permanently satisfied, i.e. if the precondition happens not to be satisfiable anymore, then the task is aborted. On the contrary, an *optional* precondition is considered as satisfied (and hence removed from the task's list of preconditions) if it is actually satisfied *or* if it happens that its own satisfiability becomes **unsatisfiable**. In this case, the task is not aborted. Figure 2.4 illustrates these precondition mechanisms.
  - **Very Urgent Task (VUT) mode:** This mode is a way to trigger a priority task, preventing any incompatible task to be executed during this time: the list of incompatible tasks to prevent should be provided as parameters of the task insertion. If an incompatible task is already running, it is interrupted. Otherwise, if an incompatible task is scheduled, then it can be either canceled (and de-scheduled) or only delayed (its preconditions are updated taking into account the task being inserted in VUT mode). The expected effect on scheduled incompatible tasks should be specified as well in the parameters of the task being inserted.
  - **DEPendant (DEP) mode:** It is a shortcut to insert a task with as many preconditions as tasks currently scheduled: each precondition is satisfied when the corresponding task triggers its "end of task" event. Moreover, these are *mandatory* preconditions (i.e. as defined in the SEQ insertion mode).
  - **Non Urgent Task (NUT) mode:** It is also a shortcut to insert a task, setting as many preconditions as tasks currently scheduled: each precondition is satisfied when the corresponding task triggers its "end of task" event. However, contrary to the DEP mode, these are *optional* preconditions (i.e. as defined in the SEQ insertion mode).
- **dynamic tasks aborting:** This mechanism enables the possibility to request tasks abortion in the current plan. If the task is already running, then the abortion of the task is an interruption. If the task is not yet running, then the abortion is a cancellation (the task is de-scheduled). The abortion triggers a propagation mechanism, that checks which of the scheduled tasks depend on the aborted task (i.e. the tasks having a precondition expecting an event from the aborted task, like a "end-of-execution" event): if the dependence is

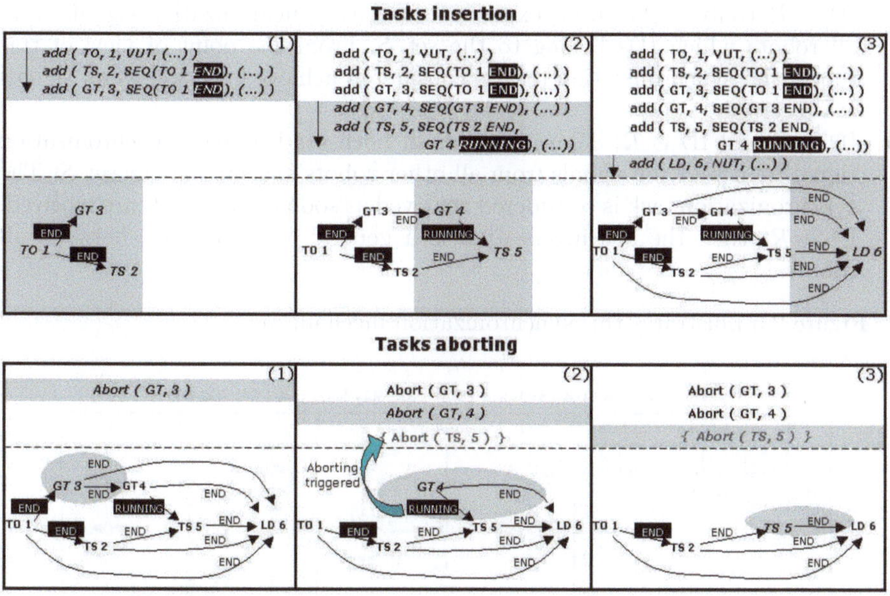

Fig. 2.4. Top: Examples of tasks insertion and illustration of the corresponding preconditions dependencies. (1): a VUT task and SEQ tasks with single mandatory precondition. (2): SEQ tasks with both mandatory and optional preconditions. (3): NUT task. **Bottom:** Examples of tasks aborting (1) and illustration of aborting propagation to dependent tasks having mandatory preconditions (2) and (3).

a *mandatory* precondition, then this task is also aborted, and so on. If the dependence is an *optional* precondition, then the dependence is removed as if the precondition was satisfied, and the corresponding task is not aborted.

The level 3 of decisional autonomy introduces an additional mechanism intended to enable autonomous synchronizations from different robots MLEs. A synchronization can be requested to a given MLE as a particular task, that produces events (start, running, end...) in the same way as usual tasks do. It is also possible to insert a synchronization task with particular insertion modes as defined previously. Two "roles" are specified as parameters of a synchronization task: sender ($\mathcal{S}$), and receiver ($\mathcal{R}$): $\mathcal{S}$ and $\mathcal{R}$ are the set of robots considered respectively as senders and receivers of the synchronization. When a synchronization task is processed, the MLE checks whether its own ID is noticed in the $\mathcal{S}$ or $\mathcal{R}$ sets. Three situations may occur:

- ID $\in \mathcal{S}$ (only): The MLE has to send a synchronization signal to all robots which ID belongs to set $\mathcal{R}$. This signal contains the synchronization task's ID, and also this robot's ID. From this robot point of view, the task is considered achieved.

- ID $\in \mathcal{R}$ (only): The robot expects to receive synchronization signals from all robots which IDs belong to the set $\mathcal{S}$. From the point of view of this robot, the synchronization task is considered achieved as soon as all signals are received.
- ID $\in \mathcal{S}$ and ID $\in \mathcal{R}$: The robot should both send its own synchronization signal then wait for signals from all other robots specified in the set $\mathcal{S}$. The synchronization task is considered achieved as soon as all signals are received. If $\mathcal{S}$=$\mathcal{R}$, then the synchronization is a general "rendez-vous" between all robots.

Figure 2.5 illustrates this synchronization mechanisms.

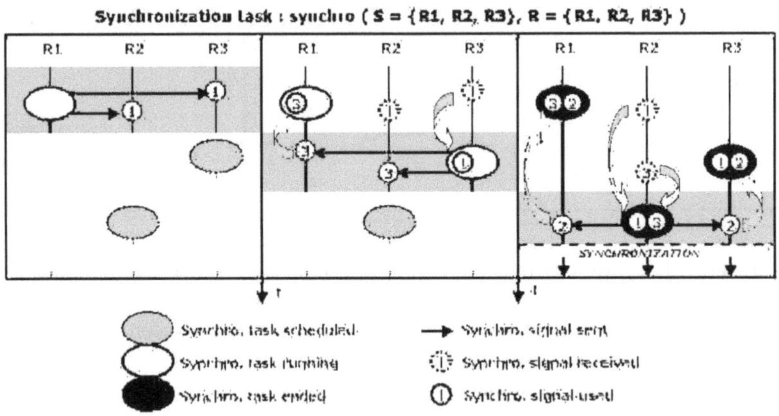

Fig. 2.5. Illustration of a synchronization task with 3 robots, in the case of a general "rendez-vous"

2.3.3 Illustration

These various mechanisms have been instanciated within the COMETS project [18], and exploited in a scenario that implied a fire detection, confirmation and monitoring task, plus a mapping task. A video depicting the various phases of the scenario can be seen at [1].

2.4 Multi-UAV Distributed Mission-Planning

2.4.1 General Considerations Related to the Planning Scheme

The symbolic planner used is based on the Shop 2 HTN planner [28], exploiting a hierarchical definition of the planning domain. According to this paradigm, high level *methods* are decomposed into lower level tasks (either other methods or operators) when methods' preconditions are satisfied, until the planner reaches primitive tasks (*operators*).

We introduce time thanks to a particular encoding of the domain based on *Multi-Timeline Preprocessing* (MTL), according to [28]. This scheme enables to express durative and concurrent actions, which is very relevant in robot's tasks planning.

Moreover, we allow, for every task, the possibility to deal with temporal constraints: these time constraints are related to wishes or requirements, expressed in missions requests. Four possible time constraints are enabled in this way: *start before, start after, end before, end after.* When a method generates sub-tasks during its decomposition, these sub-tasks inherit the time constraints.

We distinguish two kinds of operators: *actual operators* (AO), corresponding to explicit tasks in the generated plan, and *convenience operators* (CO), manipulating intermediary data, but not directly dealing with actual robot's tasks.

AOs have the following properties:

- An unique ID (generated during the planning process).
- A dependence list: dependencies dealing with other (previous) operators. This list built using the MTL properties is used when the MLE receives a plan to execute: the dependencies are then turned into preconditions.
- A relative starting time: a time interval where the task's starting should be triggered.
- A duration: provided by the *specialized refiners.*
- Time constraints, inherited from higher level methods decomposition, during planning.
- Some parameters, according to the operation's type.

These AOs mainly match the elementary tasks defined previously (e.g take-off, gotoXYZ, etc.). AOs may also match highest level tasks which can not be refined in the only UAV's context : such tasks require multi-UAV refinements, which occur in a second step, trough the *interactions manager.* The duration of such a *Joint Task* (JT) is not necessarily relevant during plan building, since it may depend on the task refinement issue in the multi-UAV context: in this case, the duration is let "unknown" for this task.

On the other hand, the COs are related to intermediary operations, such as calling the specialized refiners during planning. Applying such a CO operator is required before applying any AO operator, since it provides a way to link symbolic knowledge with actual models of the world: environment, UAVs, communication, etc.

2.4.2 Exploiting the Specialized Refiners During the Planning Process

Figures 2.6, 2.7 and 2.8, illustrate a "gotoXYZ" method (Fig. 2.6) giving rise first to the computation (CO, Fig. 2.7) of data related to "gotoXYZ" task, then applying the primitive "gotoXYZ" task (AO, Fig. 2.8). The "compute-gotoXYZ" operator sends a request to the specialized refiners for the refinement of the "gotoXYZ" task, taking into account initial location and destination location,

```
(:method (general-gotoxyz ?destloc ?time-constraints)
  ; preconditions
      ((uavloc flying ?startloc) (not (eval (eql '?startloc '?destloc))))
  ; subtasks
      (:ordered (!compute-gotoxyz ?startloc ?destloc)
                (!task-gotoxyz ?id ?dependences ?startloc ?destloc
                         ?waypoints ?start ?duration ?time-constraints)))
)
```

Fig. 2.6. A Shop method for the generation of a "gotoXYZ" primitive

```
(:operator (!compute-gotoxyz ?startloc ?destloc)
  ; preconditions
      ((assign ?result (compute-data 'gotoxyz (list '?startloc '?destloc))))
  ; delete list
      ()
  ; add list
      ((eval-ok gotoxyz ?result))
  ; cost
      0
)
```

Fig. 2.7. CO example: calls the specialized refiners features

```
(:operator (!task-gotoxyz ?currentid ?dependences ?startloc ?destloc
            ?waypoints ?start ?duration ?time-constraints)
  ; preconditions
    (
      (eval-ok gotoxyz ?pre-computed-data) ; (1)
      (assign ?duration (get-duration '?pre-computed-data)) ; (2)
      (assign ?waypoints(get-waypoints '?pre-computed-data)) ; (3)
      ...
  ; delete list
      ...
  ; add list
      ...
  ; cost
    (get-cost '?pre-computed-data)) ; (4)
```

Fig. 2.8. AO example: the "gotoXYZ" operator

and the returned result is added in the current planning state (through the logical atom "eval-ok...", in the operator's "add list" field). Then the "gotoXYZ" operator exploits the corresponding result (line (1) on Fig. 2.8). Finally, the result is parsed into the different relevant data, e.g. duration, waypoints and costs associated to the "gotoXYZ" operation application (resp. lines (2), (3) and (4) on Fig. 2.8).

```
(TASKREQ
   TASK-GOTOXYZ 13
   (DEPENDENCES ((ENDED 12) (ENDED 11)))
   (PARAMS
      (WAYPOINTS ((WP 100.000000 -140.000000 100.000000 0 0 0 0 -1)
         (WP 100.000000 -130.000000 90.000000 14 1 -1 1 -1)
                      :
         (WP 120.000000 -10.000000 50.000000 164 13 -1 20 -1)))
     (START-TIME 240 352.000000)
     (DURATION 88 113.000000)
     (TIME-CONSTRAINTS NIL NIL NIL NIL)
)
```

Fig. 2.9. "GotoXYZ" task, ready to be executed

Figure 2.9 illustrates an instance of "gotoXYZ" task, as it appears in a final plan.

Actually, the specialized refiners have the means to process data for much more complex tasks, such as tasks requiring both refinements for perceptions and path planning (e.g. TSP with planned perceptions, see Sect. 2.4.3).

Exploiting Resulting Plans - Multi-UAV Coordination Issues

Only the AOs are notified in the final plan. Such a plan is ready to be executed **if** it does not contain any task requiring coordination with other UAVs, i.e. JTs. However, if the plan contains JTs, the plan coordination is performed in a second step, through the *interaction manager*.

The interaction manager provides the means to coordinate UAVs activities, relying on distributed negotiation mechanisms. All the tasks requiring multi-UAV interactions (simple synchronization or more complex JTs) are processed in the interactions manager, so that the joint operations can be coordinated, for each involved UAV, in terms of space and time.

Detail related to the interactions manager is not provided here, since it is still ongoing work. Mainly three issues are tackled:

- Temporal coordination: achieved relying on UAVs synchronizations. We defined and implemented a scheme to enable incremental negotiations related to possible time intervals synchronization. As a result, a group of UAVs acknowledge a common time interval in which the synchronization should occur.
- Spatial coordination: we consider interactions models, to reason about the interactions requirements within the JTs. Afterwards, during plan execution, collision avoidance can be safely achieved applying a Plan Merging Protocol [3] on the planned trajectories of UAVs.

- Tasks re-allocations: this issue consists in enhancing the global activity of the UAVs, allowing them to re-distribute some tasks, when relevant. For each UAV, the relevance should be assessed w.r.t. the current tasks costs / utility in the current plan (see Sect. 2.5).

During coordination, the interaction manager may as well request computations / refinements related to the environment and UAVs models, i.e. relying on the *specialized refiners.*

As a result of these coordination processes, a coordinated, ready-to-be-executed (but not necessarily definite) sequence of tasks is provided and inserted in the current MLE's plan.

2.4.3 The Specialized Refiners Toolbox: Overview

The specialized refiners provide a wide set of features to support tasks decompositions and refinements during planning and coordination. They rely on different models (environment, UAV, etc) regularly updated during the UAV's activity, and offer (through a common interface) a set of services related to paths generation, perception planning and communication constraints satisfaction checking. These different processes are performed in a timely manner, so that the symbolic planner may use them in a transparent way during plan building.

The main point here is to provide the planner and the interaction manager with information that will allow them to estimate the ability of the robot to perform a given task in a given (dynamic) context, to compute the various costs and to weight the consequences of inserting a given task into the current robot plan. Hence such information should be sufficiently realistic and produced in an efficient way in order to be used on-line. Indeed, the overall process is incremental and is subject to frequent revisions.

Models

The environment model developed provides two kinds of information: ground surface data and airspace data. The ground model is a 2D array composed of square cells whose attributes are related to fire, mapping and alarm monitoring. A burning factor representing the burning risk is associated to each cell. The airspace model is represented by a 3D array of voxels, and gives relevant information for trajectories and perceptions planning. It indicates whether a voxel is free (and safe) for flying or not. Moreover, considering the communication model (described hereafter), we are aware of the voxel's communication coverage (regarding a control center for instance). Potential waypoints for trajectories planning are nodes located at the center of voxels' facets. A single voxel has 6 nodes and shares each of them with a neighbor. Nodes are connected by edges. Each edge is labeled with the cost for the UAV to move from one node to another adjacent node.

We also use a generic UAV model that provides information concerning flight capabilities and available resources. It is mainly used for flight time estimation purposes; it also gives information about possible orientation for perception devices. The perception model contains technical characteristics related to perception devices (e.g. expected coverage...), and informs about the availability of the sensors. Finally, we have also implemented a (quite simple) communication model that allows to estimate the ability to communicate between two entities. With omni-directional antennas and no solid obstacle between the sender and the receiver, the following conditions must be satisfied:

$$\frac{\text{power}_{\text{sender}}}{4\pi \left(\text{distance}_{\text{sender/receiver}}\right)^2} > \text{sensibility}_{\text{receiver}} \tag{2.1}$$

According to these models, various "services" can be provided: the next section provides algorithmic details related to these features.

Algorithms

Simple path planning

Path planning is performed here, in a simple way (A* based), in order to compute a path in a discretized 3D environment (Sect. 2.4.3). The planner takes into account obstacles (hilly ground) and no-flying zones (see Fig. 2.10).

This planner is used to compute all possible trajectories between all potential robot's mission objectives. We use an extension of this scheme to find the shortest

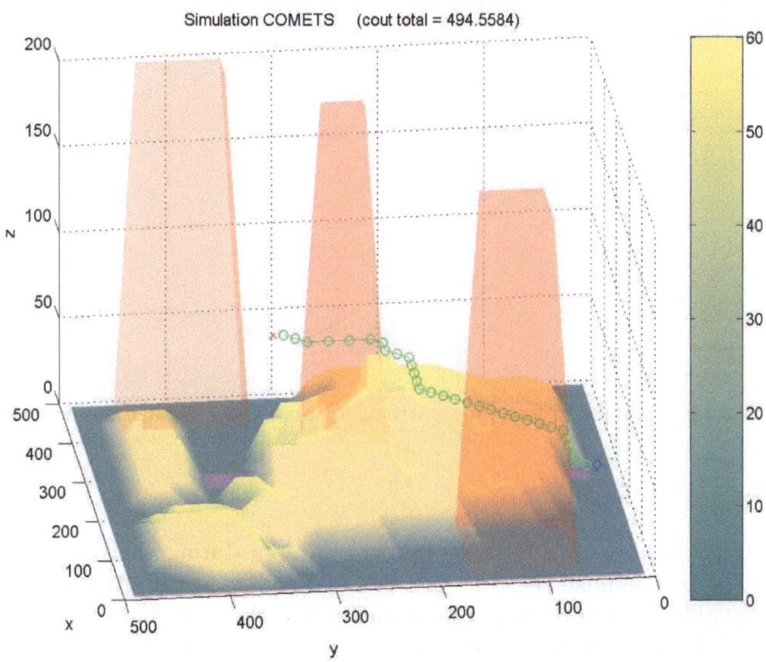

Fig. 2.10. Simulation for path finding

path between several points (i.e. Traveling Salesman Problem (TSP)). This is used by the planner and the interaction manager to compute the best way to insert a new task and to estimate the cost of its insertion in the robot's plan.

The TSP is approximated here with a simple stochastic algorithm, using two operations: insertion and permutation. At each iteration, a local *minimum* is found from an initial random solution. After a given number of iterations the best result is kept as the global solution. The number of iterations has been chosen experimentally, as a trade-off between the quality of the solution and the computation time. The algorithm gives quite homogeneous results (as far as the number of points is not less than 30, which is largely enough for a complete UAV mission).

Mapping

The goal of the mapping task is to cover a whole given area in the shortest time. In this problem, we try to minimize the number of turns, according to [27], where the authors introduce an efficient method of area decomposition for multi-UAV contexts and a relevant way to apply sweeping patterns. Turning is considered as critical because an UAV may slow down as its direction is changing. Moreover, trajectories are more difficult to follow in turns, hence perception cannot be achieved as efficiently as in straight lines. The principle of the mapping algorithm is to select a favored direction (along the longest straight line inside the area), and then to apply a sweeping pattern considering this direction, as shown on Fig. 2.11. We assume that areas are (or can be divided) into convex polygons.

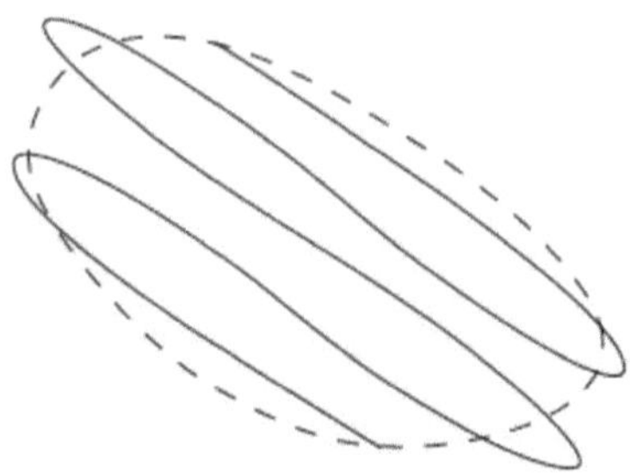

Fig. 2.11. Example of sweeping pattern, for mapping/coverage applications

Detection

This activity requires the UAV to fly over an area during a given amount of time, trying to minimize the time between two flights over a given ground cell. Moreover, different priorities can be associated to the cells. For instance in the context of the COMETS project, dry vegetation is more likely to burn than water pools: detection activity should be performed with respect to terrain's burnability.

We propose an algorithm based on potential fields. Each cell of the ground is associated to a point of the field, initiated to its maximum, and decreasing with the time according to the equation:

$$P = \mathrm{e}^{-r \cdot \Delta T}, \tag{2.2}$$

where P is the potential of the considered point, $r \in [0, 1]$ is the risk factor and ΔT is the time since the last visit. Perception is not limited to a single cell, depending on the perception device's aperture and flying altitude. At each step of time, ΔT is incremented according to the factor

$$1 - \mathrm{e}^{-\left(\frac{(x-x_u)^2+(y-y_u)^2}{\sigma^2}\right)}, \tag{2.3}$$

where (x, y) is the location of a cell, (x_u, y_u) is the location of the UAV and σ is a parameter representing the sensor's aperture (and so its influence on adjacent cells). After each increment, the value ΔT of the cell located under the UAV is set to zero, corresponding to the maximum of P. The next move follows the steepest gradient in the potential field. Figure 2.12 shows the state of the potential field during a simulation.

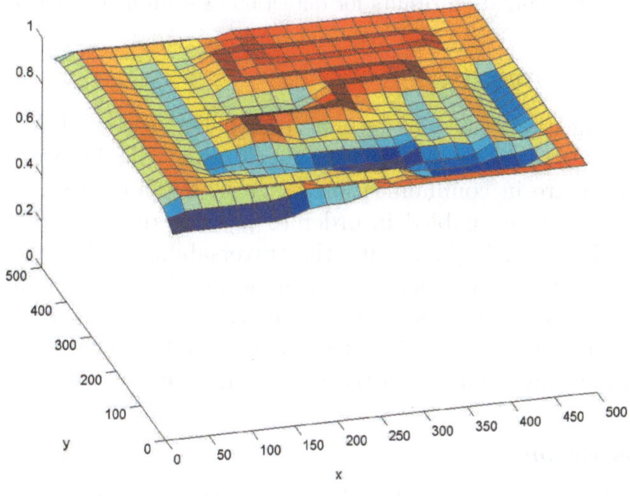

Fig. 2.12. Example of potential field for detection applications

Figure 2.13 provides an example of path followed by an UAV over an area with heterogeneous parts : high risk parts clearly appear with a large number of flights, whereas low risk areas are rarely visited. Even for very low risk areas, the potential slowly decreases until reaching a lower value than high risk areas, and hence also attracts the UAV after a given lapse of time. As a consequence, except if r is equal to zero, every cells are explored at least once, after a long enough time.

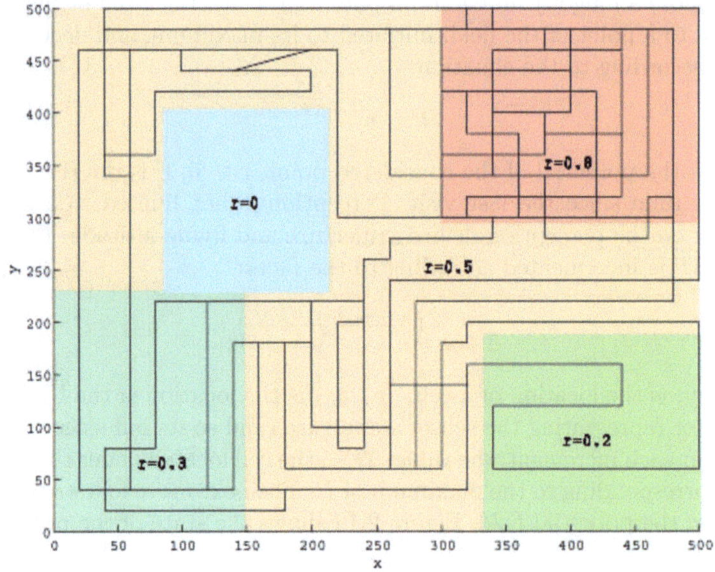

Fig. 2.13. Simulation results for detection over an heterogeneous area

Constraining basic requests

When communications need to be maintained between an UAV and another entity (e.g. other UAV or control center...) during a flight, we have to check if related voxels are in communication range with the entities. If it is not the case, a "filter" can be enabled in order to prevent the path planner to build paths across these voxels (increasing the traversability cost of these voxels, for instance). Other types of constraints can be applied through basic requests, such as a *preliminary path computation*: the refiner should take into account a preliminary path, before to perform the requested refinement. Indeed, the initial location can have an influence on the refinement issue.

2.4.4 Illustration

This section illustrates the developed planning scheme in the context of scenario that involves three UAVs.

Mission and Scenario

Mission

The general mission's goal is to perform fire detection and fire monitoring over a given area A. The initial task allocation is performed by a human operator. Fire alarm detection should be performed by one UAV over A. Every located alarm should be confirmed by another UAV. If an alarm is confirmed, then 2 UAVs should perform coordinated monitoring around the fire.

Scenario

Three UAVs are introduced in this scenario: one blimp (K), not very maneuverable but well adapted to high altitude flights, and two copters (H and M), much more maneuverable, having hovering capabilities, well adapted to low altitude flight.

K is requested to perform detection over A. After a certain amount of time, a first fire alarm is raised over the location L1, then a second fire alarm is raised over the location L2. H is requested to make perceptions around L1, and M should make perceptions around L2. In L1, the alarm is infirmed (false alarm). In L2, the alarm is confirmed: H is requested to perform coordinated perceptions with M around L2, for monitoring purposes (requires a synchronization of the monitoring). During this time, K keeps on performing fire detection around L1 and L2. The monitoring activities performed by H and M should go on until K's detection activity is ended. After a certain amount of time, K stops its detection activity: a synchronization signal is sent to H and M. All the UAVs come back to the base station.

Running the Scenario

Requests hereafter deal with high level Shop methods: once requested to Shop, they are decomposed into refined elementary tasks (resulting UAVs' refined plans are illustrated on Fig. 2.14), exploiting the specialized refiners abilities. The sweeping pattern for fire detection is computed by the specialized refiners, as well as the most fitted perception locations close to L1 and L2 (for H and M), maximizing the perception utility (Fig. 2.15 depicts a simulated instance of this scenario).

K blimp's mission

- K should perform detection over A during 15 minutes;
- THEN K should send sync.signal (S1) to H and M.
- THEN K should come back to the base station.

On Fig. 2.14, task 11 is a "goto" task leading to area A. Task 12 is a "goto-list" task associated to the detection pattern computed by the specialized refiners. As task 12 is running, the perceptions are simultaneously triggered (task 13). Then once the synchronization is achieved, the "goto" task 15 makes K come back to the base station.

H copter's mission (part 1: H1)

L1 alarm raised (through K's perceptions): should be confirmed by H.

- H should make perceptions in L1 during 1 minute.
- THEN H should wait for further orders in secure mode.

Task 21 (Fig. 2.14) is a "goto" task leading to L1.

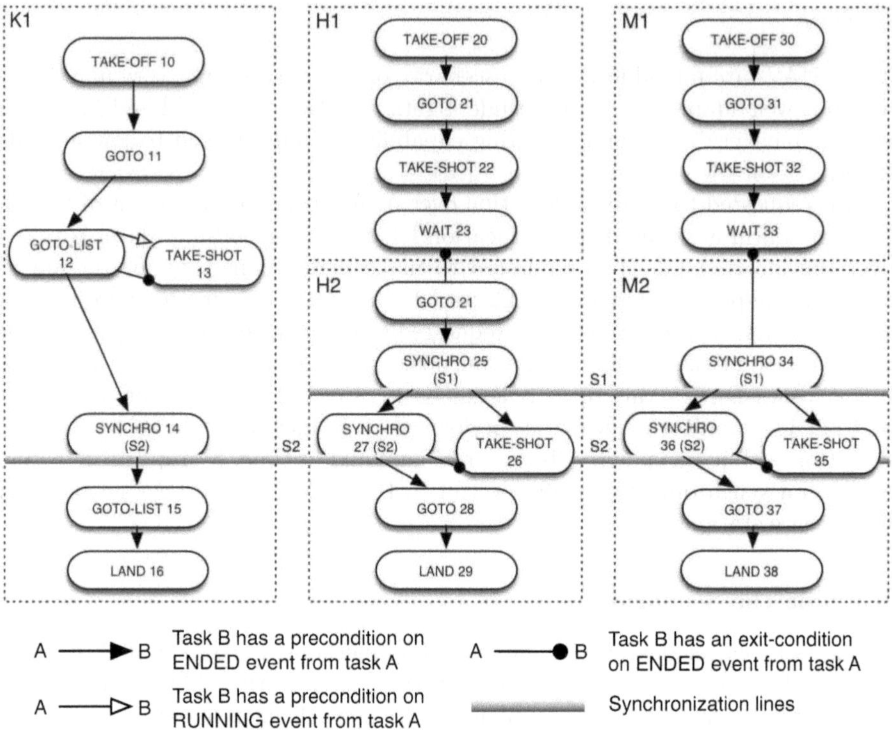

Fig. 2.14. Example of scenario involving 3 UAVs: K,H and M's plans

M copter's mission (part 1: M1)

L2 alarm raised (through K's perception): should be confirmed by M.

- M should perform perceptions in L2 during 1 minute.
- THEN M should wait for further orders in secure mode.

Task 31 (Fig. 2.14) is a "goto" task leading to L2.

M copter's mission (part 2: M2)

L2 confirmed (M's perception): should perform coordinated monitoring.

- M should perform monitoring activity of L2 with H until receiving synchronization signal from K.
- THEN M should come back to the base station.

On Fig. 2.14, task 34 is the synchronization with H for monitoring (task 35). Task 36 is the synchronization with K, which achievement stands as exit condition for task 35. Then task 37 is the "goto" task back to the base station.

H copter's mission (part 2: H2)

L1 alarm is wrong, and L2 is confirmed (through M's perception data processing): should perform coordinated monitoring.

- H should perform monitoring activity of L2 with M until receiving synchronization signal from K
- THEN M should come back to the base station.

On Fig. 2.14, task 24 is a "goto" task leading to L2. Then task 25 is the synchronization with M for monitoring (task 26). Task 27 is the synchronization with K, which achievement stands as exit condition for task 26. Then task 28 is the "goto" task back to the base station.

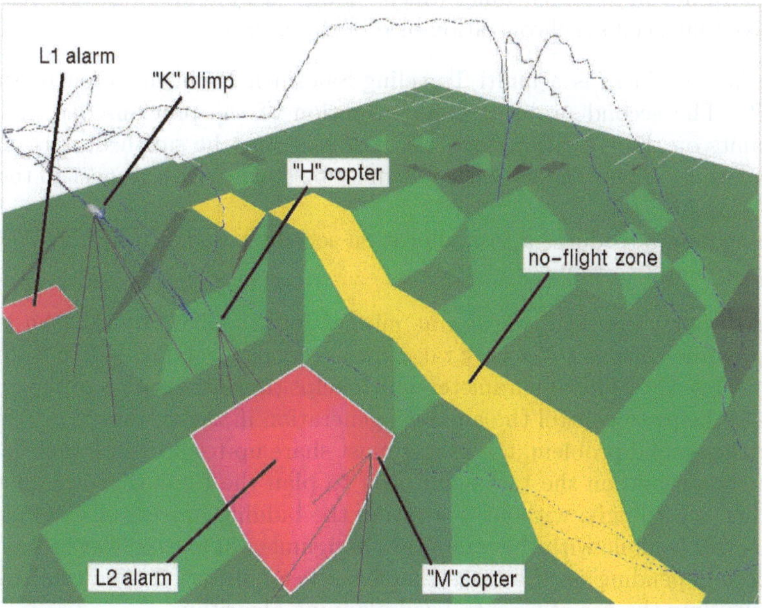

Fig. 2.15. Level 4 features in simulation: coordinated monitoring over L2

2.5 Distributed Task Allocation

Problem Statement

The focus in this section is put on the distributed task allocation issue, an ability the level 5 robots must be endowed with. Given a system of several robots, and given a mission defined as a a partially ordered set of tasks, we want the robots to allocate all the tasks to each other and build their plans accordingly in order to complete the mission. They should also be able to dynamically modify the allocation, and consequently their plans, to adapt to changes in the environment or to new requests issued by the operator. The system must also satisfy constraints on energy resources and communication ranges, that are both limited. Also, all the robots must ensure sufficient energy to go back to their starting point when their tasks are achieved.

In the context of an environment surveillance mission, the tasks the system of robots must achieve consist of (*i*) navigation tasks (i.e. reach a given position), and (*ii*) perception tasks. The latter tasks can be achieved while the UAV hovers at a given position, or while it follows a predefined path (*e.g* circling around a given target), but they can also imply the simultaneous presence of several UAVs, hovering or moving according to predefined geometric patterns. Therefore, two distinct problems must be solved:

1. allocation and planning of navigation tasks,
2. constraints on tasks schedules which enable the system to deal with complex tasks that need synchronization of robots activities.

The first problem is a multi Traveling Salesmen Problem (often referred as *m-TSP*). The second problem is an extension of the previous one, in which constraints on the execution dates of the tasks must be satisfied: it is possible to stipulate for a task the date when it must be started with respect to the start date of another task.

This section introduces a task allocation scheme based on *Contract-Net*, that is innovative in two ways :

- For the first problem, we aim at minimizing a global criteria (the longest trip) while *Contract-Net* only takes into account local data. We introduce in *Contract-Net* a global parameter which helps the optimization of the criteria. It also helps to control the auctions generation in the system.
- For the second problem, the robots must share up-to-date data that describe the constraints on the tasks and need to plan the tasks that are linked together accordingly, without disrupting the bidding process of *Contract-Net*. This will be done with the temporary assignment of master/slave roles to the robots depending on the tasks that have been allocated. The challenge lies in avoiding the use of a centralized planning algorithm.

Related Work

Rabideau et al. made a comparison of several methods for tasks allocation in [11]. They emphasize algorithms with three degrees of distribution, the most distributed one is based on the *Contract-Net* protocol. In [7] Bellingham et al. successfully implement in simulation an algorithm for the *optimal fleet coordination problem.* Their algorithm does not address the problem of synchronization between tasks and can be classified in level 2 from [11]. Dias and Stenz also studied various approaches to the task allocation problem with multiple robots [14] and came to the point that distributed algorithms based on *Contract-Net* suit the needs. Note that a number of distributed schemes for task allocation in multi-robot domains have been proposed in the literature, being [5] one of the firsts. ALLIANCE [29] is a distributed behavior-based architecture, which uses motivations that enable/inhibit behaviors, resulting in tasks (re)allocation.

Contract-Net has been introduced by Smith in [34] and further developed by Sandholm [31]. Since 1999, *Contract-Net* has been widely used in multi-robot applications [36, 13, 20]. Stenz and Dias worked on an architecture called *TraderBots* in which *leaders* can optimize the plan of several other robots [15, 16], Matarić explored various strategies for the *Contract-Net* protocol in [26]. Several studies dealing with concrete missions such as buildings or planetary exploration [25, 42, 12] or emergency handling [26], have shown the feasibility and the performance of the *Contract-Net* architecture in real world situations.

To our knowledge, on the problem of allocation and planning of non-independent tasks in a distributed multi-robot system, only one paper from Kalra and Stentz [22] presents preliminary results on the *sweeping perimeter problem*. In this work, the temporal window taken into account is very small, the coordination is explicit between a limited number of robots (one robot and his two neighbors), and the market-based approach is not fully exploited since the auctions imply three agents only.

The next section introduces an *equity coefficient* that is used in the bids evaluation and to control the auction generation process within *Contract-Net*. Quantitative simulation results obtained on the m-TSP problem illustrate the improvements brought by the consideration of this coefficient with respect to a plain *Contract-Net* approach. Section 2.5.2 deals with the introduction of time constrained tasks. It shows that the introduction of simple execution date constraints can help to cope with cooperative tasks, that are either requested by an operator or automatically generated within the system, to establish communication relays for instance.

2.5.1 Contract-Net with Equity

In the classic market-based approach, each agent (for us robots) can make a public auction for one of its tasks, and then the other robots can bid on that task using a given cost function. The winner of the bidding process gets the task and must insert it in its plan. In order to drive the process towards an optimal solution, one agent can sell a task only if the bidden cost for the execution of that task is at least less than a certain amount of its own execution cost (generally 10% less). The cost function we will use here is simple and is calculated from the distance the robot will travel.

Equity Factor

The aim is to obtain an allocation that minimizes the length of the longest trip, which also can be seen as minimizing the duration of the mission. Our idea is to address this global optimization problem by considering two aspects: first, *Contract-Net* is used to assign tasks to the robots at a low cost so as to keep the total distance traveled by the team of robots not too far from optimality, and second, equity is enforced between the robots so as to really distribute the tasks among them and obtain a mission which is as short as possible.

For this purpose, we introduce a measure of equity called *equity coefficient* (C_{eq}). Each robot can compute its own workload (wl) using a cost function: the workload is the cost of the whole plan of the robot. The robots broadcast the value of their workload to the others and each one can compute its C_{eq}. For the robot A the formula is:

$$C_{eq}^{A} = \frac{wl(A) - \overline{wl}}{\overline{wl}} \tag{2.4}$$

where $\overline{wl}$ is the mean of $wl(.)$ over the robots for which A knows the workload. Indeed, since we consider limited communication range, A may have only a partial knowledge of the workloads. The meaning of this coefficient is :

- $C_{eq}^{A} < 0$: robot A has a too small plan with respect to the other robots.
- $C_{eq}^{A} > 0$: robot A is overloaded with respect to the other robots.
- $C_{eq}^{A} > C_{eq}^{B}$: robot A has more work than robot B.

Equity Factor and Task Evaluation

In *Contract-Net* a task is allocated to the robot which can insert it in its plan for the lowest cost; also the robot should not be too overloaded. For that, the evaluation the robot makes for a task is modified by taking into account its C_{eq}. The utility the robot A computes for the task T_1 ($ut^{A}(T_1)$) is corrected in $ut'^{A}(T_1)$ by:

$$ut'^{A}(T_1) = ut^{A}(T_1) - C_{eq}^{A} \times |ut^{A}(T_1)|$$

This correction is applied to the utility computed by both the auctioneer and the bidder. By this mean *Contract-Net* is influenced the way we want:

- A robot with a high workload will more easily reallocate its tasks and will get new tasks with more difficulty because its utility for the tasks is lowered.
- On the contrary, a robot with a low workload will be more easily allocated new tasks but will give up its own tasks with more difficulty because its utility for the tasks is increased.

Control of the Auctions Generation

The problem here is we do not want several auctions being launched at the same time. Basically, the *Contract-Net protocol* does not provide any details when the agents of the system can start an auction, and other papers do not emphasize this point either. Our need is to keep the system entirely distributed, so we do not want an authority which would give the right to the robots in turns, and we want to keep the system dynamic so we do not want to give to each agent a static list which would define the turns for the auctions.

Our solution is inspired by the *token-ring* networks in which a token passes from one computer to another to give them the authorization to send their data over the network. Here the token allows the robot to make an auction.

Token circulation

The robot that has got the token is the auction leader. If another robot is willing to make an auction, it can ask for the token to the current auction leader. It sends its request along with its C_{eq}. The owner of the token collects all the requests, it is also allowed to request for the token. It then randomly chooses the next owner of the token, using a random distribution based on the collected C_{eq} (the more a robot is overloaded, the higher chance it has to get the token). This is done to help overloaded robots to reallocate their tasks.

Token creation

When a robot wants to make an auction, but nobody has the token, it then creates a token and uses itself to make an auction (the process is started spontaneously !). Because of communication delay, it may happen that several robots create a token at the same time, this is why we specify the following behavior:

- If a robot that is not currently an auctioneer receives several auctions at the same time, it then bids on the auction which has the higher priority *ie* the higher C_{eq} (the auctioneer gives its C_{eq} along with the auction). The other auctions are ignored.
- If a robot that is currently an auctioneer receives other auctions, it keeps on its own auction only if it has the highest C_{eq}, else it cancels its auction and can bid on the auction with the highest priority.

Results

A typical mission allocation and execution goes this way: (1) A set of tasks is given to the system (either directly by an operator, or issued from a decomposition process). (2) The base station is also a *Contract-Net* agent except that it has a high priority (an artificially high equity factor) and will never make a bid. The base starts making auctions with the tasks of the mission and goes on until all tasks are allocated to the robots. (3) The base does not keep the token any more and the robots can start making auctions. (4) The process stops when none of the robots asks for the token. A robot stops making auction when it has already auctioned all its tasks and no reallocation has occurred. If not, the robot auctions again all its tasks. This stop criterion is quite different from what has been done until now (usually a fixed number of auctions turns). (5) The mission starts being executed by the robots. The auction process starts again when new tasks are requested by the operator, or when a robot fails to achieve its plan.

In the tests, we focused on steps 2 to 4[2]. We based all our tests on the same mission: 50 points picked up in the environment have to be visited by a team of 4 robots. The points have been uniformly randomly generated once in this environment. In order to show the interest of our *equity coefficient*, we run scenarios with the coefficient disabled (we give it a fixed value so as to mimic a plain *Contract-Net protocol*). Another important point is how the robots are

[2] The results were obtained with a multi-robot simulation developed in Java.

firstly distributed in the environment. If they are scattered (the usual situation when *Contract-Net* is used) each robot is implicitly attributed a different area, the area surrounding its initial position, because of the cost function which is based on traveled distance. If they are initially grouped around a same point (which is mostly the case in operational situation), the problem is more difficult.

The results presented in Table 2.1 show that the solution obtained with the equity coefficient is improved by a factor of 2.4 over the standard *Contract-Net*. If the robots are scattered, the improvement factor drops to 1.3. This is due to the fact that the solution found by the standard protocol is already a good one. The interest of our method is that it works well even if the initial situation is not favorable. On the other side the allocation process is about 20% longer (more auctions are done) with the equity coefficient enabled.

Table 2.1. This table summarizes the statistical results over 100 runs of the simulation of four scenarios, considering a grouped or scattered start, and with or without the use of the equity coefficient. l is the length of the maximal tour, and n is the number of auctions of the allocation process.

scenario	$\bar{l}$	$\sigma(l)$	$\min(l)$	$\max(l)$	n
grouped/no equity	5248	674	3035	6150	133
grouped/equity	2195	264	1844	2954	156
scattered/no equity	2481	481	1724	4631	133
scattered/equity	1895	162	1581	2343	160

2.5.2 Time-Constrained Tasks in a Distributed Environment

The problem of constrained tasks allocation and planning for a system of multiple robots is commonly addressed with a centralized planner such as *GRAMMPS* [9]. We sketch here how we deal with simple time constrained tasks in our distributed environment.

Execution Around Date *d*

This constraint means that we will try to have a given task executed more or less at a given date. This constraint enables the system to deal with constraints on relative date of execution of several tasks expressed numerically: T_1 and T_2 simultaneously or T_1 n seconds before of after T_2.

We choose to put the constraint *execution around date d* on the tasks for several reasons:

- This constraint is *soft*, which means that there is an infinite number of solutions that satisfy it, the distributed allocation algorithm will more easily find a solution, even a bad one, and will not end to dead-lock.
- The quality of satisfaction for such a constraint is easily *measurable*, and then we are able to take into consideration this measure when we evaluate

its utility for a given robot. The quality of satisfaction for the constraint can be directly included in the bid of our *Contract-Net protocol.*

- The information needed to plan such constrained tasks is very limited and will not overload the communication bandwidth between the robots.

Constrained Tasks Tree

The constraint can be used for example to enforce simultaneous execution of two tasks. One task T_1 which is planned for execution at date d_1 puts on task T_2 the constraint *execution around date* d_1. The task T_1 is said to be the mother task, and task T_2 the child task.

Temporally, T_1 is defined by a *startDate* d_1 (the date when it can be executed at the earliest), and a *plannedStartDate* (the date when it will actually be executed). T_2 has the same attributes plus an *expectedStartDate* d_1 (the preferred date for its execution).

The allocation process must allocate T_1 before T_2 (because we need to know d_1 for bidding on T_2), but the tasks can be reallocated later. It is important to note that only T_2 is constrained, T_1 is allocated and firstly planned as usual. After the allocation process, the *master* robot R_A (the one which will be executing T_1) will choose *plannedStartDate* for both T_1 and T_2, the robot R_B (the one which will be executing T_2) is called the *slave* robot. Since the system is dynamic, changes can be made to the plans of R_A and R_B. If it happens, the slave robot only informs its master of the changes in its plan, it sends the new *startDate* for that task and then the master robot computes a new *plannedStartDate* for the execution of the two tasks which is acceptable by both R_A and R_B.

The relation master/slave between the robots is local in time (only for the execution of the considered tasks), and temporary because the tasks can be reallocated to other robots. So this is quite different from the *TraderBots* architecture [16].

Figure 2.16 presents a tree of tasks and focus on the data that are exchanged in order to plan the tasks, and Fig. 2.17 sketches the allocated tasks from the robot

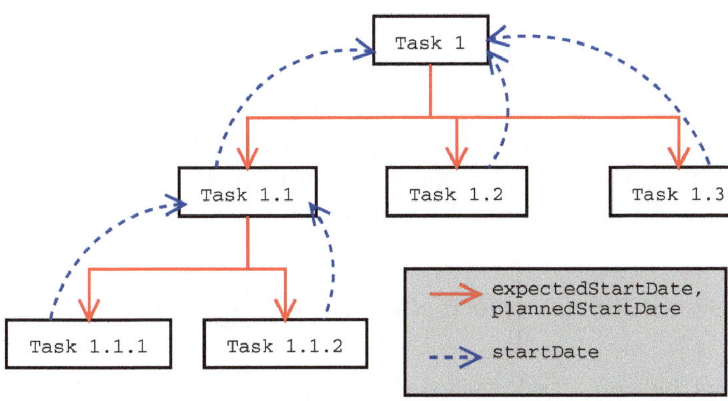

Fig. 2.16. This tree shows up the hierarchical links between tasks

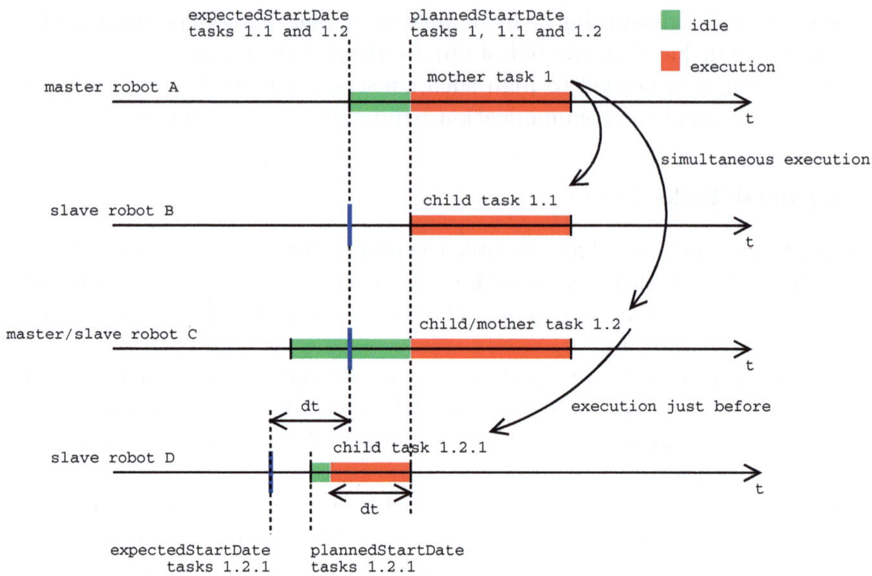

Fig. 2.17. This is an example of the plans of four robots after the allocation of four constrained tasks

point of view. The synchronization between robots is actually accomplished with the introduction of *idle* periods in the robots plans.

Evaluation of the Utility of a Task

Now the quality of satisfaction for the constraints which weight on the tasks of the plan is to be taken into account. Previously we computed the cost of a plan with its length; now we add a term for each task which reflects the constraint satisfaction quality. We call this term *deltaDate* and for the constrained task T_i the formula is:

$$deltaDate_i = |startDate_i - expectedStartDate_i| \quad + \sum_j deltaDate_{ij}$$

where $deltaDate_{ij}$ comes from the children tasks T_{ij} of task T_i. These children tasks are either allocated to the same robot or to another one.

The utility of a plan can now be computed by the formula:

$$planUtility = -\left[movingCost + k \times \sum_{task_i \in plan} deltaDate_i \right]$$

The robot bids on a task with the value $(planUtility' - planUtility)$ where $planUtility$ and $planUtility'$ are respectively the utility of the plan *before* and *after* the insertion of the task.

The factor k is here to normalize the sum. In fact we add two quantities $movingCost$ and $deltaDate$ which are not of the same nature. This becomes false if the $movingCost$ is computed with the time needed by the robots to go from one point to another. One can understand k as a scale factor, $k = 0.1$ (is the typical value we use) means that we find the periods when the robot is idle 10 times less important than when the robot is active.

Time Consistency of the Plans

We must ensure that the time constrained tasks are planned correctly to prevent the system from ending into deadlock. Here again we use a very basic planner, not really efficient, but very easy to implement and which clearly maintains consistency of the plans. Each child task is tagged with an $expectedStartDate$ and the planner will insert the task into the plan so as to respect the local chronology between children tasks of this plan.

When a modification occurs in the plan, we use a simple but rough process to maintain this local consistency: if two children tasks are not in the chronological order, they are swapped.

Assuming that the plans are incrementally built, the local consistency ensures the global consistency. Indeed, the synchronization is reached by inserting idle periods into the plans of the robots so a robot waits even a long time for synchronization rather than trying to swap tasks.

Implemented Tasks

Two new types of task have been implemented in our simulation to illustrate constrained tasks: *watchout* and *com-relay*.

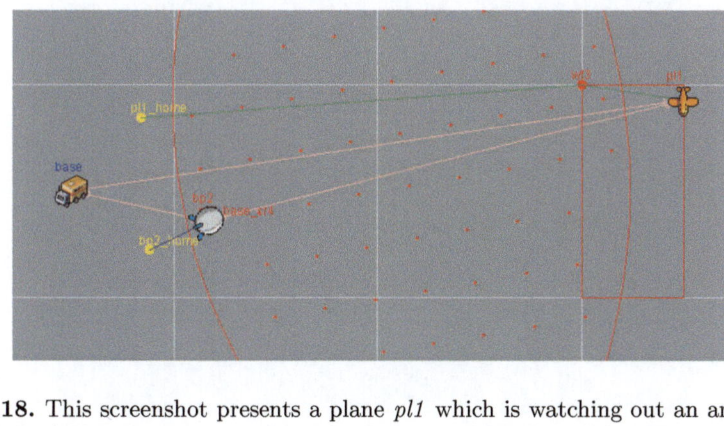

Fig. 2.18. This screenshot presents a plane *pl1* which is watching out an area and a blimp *bp2* which is a communication relay between *pl1* and the base. The communication link *pl1*→base is not available). The circle arcs enclose the area where the blimp can serve as a communication relay, and the dots represent the discrete positions (the planner of the blimp has chosen one of these positions).

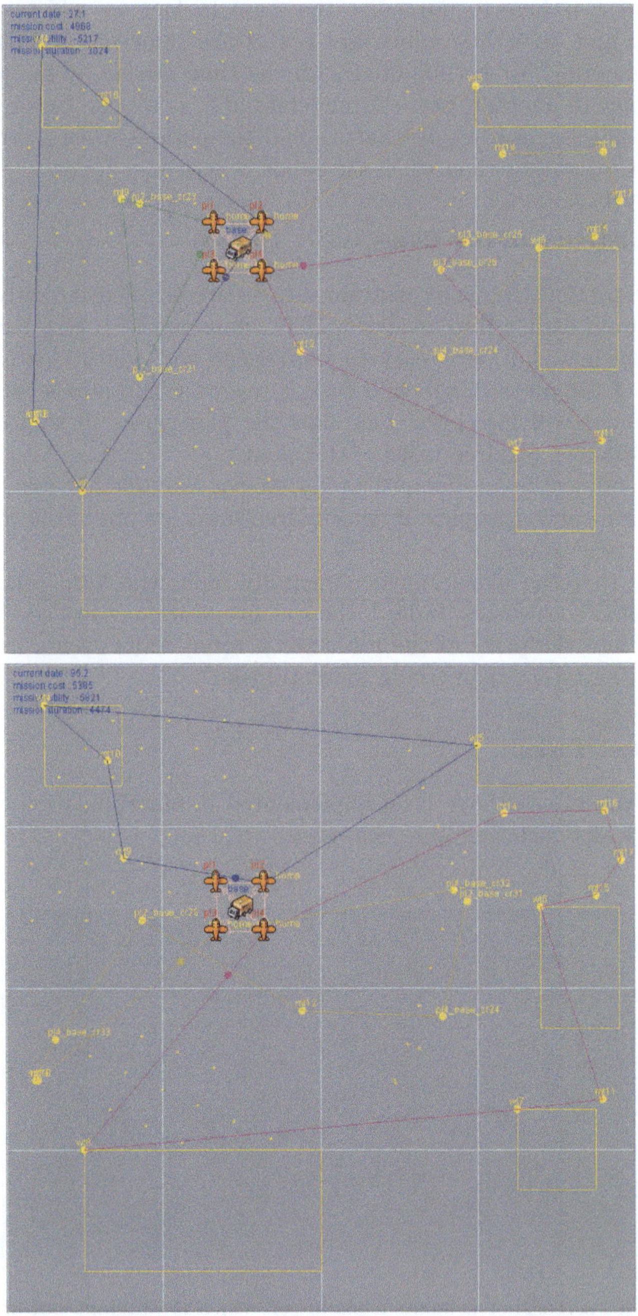

Fig. 2.19. Top: some watch-out and goto tasks are allocated to the team of robots. Bottom: The plane *pl4* (top-left) has fallen out of order and the tasks have been reallocated to the remaining robots.

The *watch-out* task consists for the robot to travel around a rectangular area to be monitored and the robot must keep communication with the base. If the communication link between the base and the robot cannot be maintained during the execution of the task, then the robot should generate a *com-relay* task between it and the base which is to be executed by another robot at the same time the watch-out task is executed. The com-relay task can be recursive, which means that several robots can be needed to effectively maintain communication between the base and the robot which will be watching out the area. Figure 2.18 illustrates these two tasks.

Illustrations

Here the results are more qualitative. The simulator shows that even with very simple planning algorithms in a distributed environment, we manage to allocate and plan a mission correctly. Figure 2.19 presents what we obtain with our simulator on some examples. It illustrates the strategy found by the team: since for the watch-out tasks two robots are needed (one for the given task, and one more for the com-relay), we can see that the four robots are splitted into two teams of two robots and each team takes care of a part of the environment. When there remains only three robots, the solution is more complex and less structured, but is valid and does not appear to be very sub-optimal.

2.6 Summary

This chapter provided insights on the definition of the architecture and associated algorithms to allow the deployment of a fleet of heterogeneous UAVs. It introduces five levels of autonomy for an UAV integrated within a multi-robot system, and proposes algorithms to fulfill three main decision-making functionalities: an executive system common for all autonomy levels that handles the coordination and task execution issues, a planning scheme based on a Hierarchical Task Networks planner completed with plan-refiners that consider the actual domain models, and an instance of the contract-net protocol that can handle task allocation for complex multi-UAV missions.

Further developments should however be made towards the overall integration of the various concepts and algorithms. For instance, the planner used in the task allocation process to insert the tasks into the plans of the robots is very simple, and the refiners could advantageously be used instead.

References

1. Final demonstration of the COMETS project. `http://www.laas.fr/~simon/eden/gallery/videos.php`.
2. R. Alami, R. Chatila, S. Fleury, M. Ghallab, and F. Ingrand. An architecture for autonomy. *International Journal of Robotics Research*, 17:315–337, 1998.

3. R. Alami, F. Ingrand, and S. Qutub. A scheme for coordinating multi-robot planning activities and plans execution. In *Proceedings of the European Conference on Artificial Intelligence*, 1998.
4. R. Arkin. Motor schema-base mobile robot navigation. *International Journal of Robotics Research*, 1990.
5. H. Asama and K. Ozaki. Negotiation between multiple mobile robots and an environment manager. In *Proceedings of IEEE Int. Conf. on Robotics and Automation*, pages 533–538, Pisa, Italy, 1991.
6. T. Balch and R. Arkin. Behavior-based formation control for multirobot teams. *IEEE Transactions on Robotics and Automation*, 14(6):926–939, 1998.
7. J. Bellingham, M. Tillerson, A. Richards, and J. P. How. Multi-task allocation and path planning for cooperating UAVs. In *Proceedings of the Conference on Cooperative Control and Optimization*, 2001.
8. R. Brooks. A robust layered control system for a mobile robot. *IEEE Journal of Robotics and Automation*, 2(1):14–23, 1986.
9. B. L. Brumitt and A. Stentz. GRAMMPS: A generalized mission planner for multiple mobile robots in unstructured environments. In *Proceedings of the IEEE International Conference on Robotics and Automation*, pages 1564–1571. Piscataway: IEEE Computer Society, May 1998.
10. L. Buzogany, M. Pachter, and J. D'Azzo. Automated control of aircraft in formation flight. In *Proceedings of the AIAA Guidance, Navigation and Control Conference*, pages 1349–1370, Monterey, USA, 1993.
11. S. Chien, A. Barrett, T. Estlin, and G. Rabideau. A comparison of coordinated planning methods for cooperating rovers. In C. Sierra, M. Gini, and J. S. Rosenschein, editors, *Proceedings of the Fourth International Conference on Autonomous Agents*, pages 100–101, Barcelona, Spain, 2000. ACM Press. poster announcement.
12. M. B. Dias, D. Goldberg, and A. T. Stentz. Market-based multirobot coordination for complex space applications. In *Proceedings of the 7th International Symposium on Artificial Intelligence, Robotics and Automation in Space*, May 2003.
13. M. B. Dias and A. T. Stentz. A free market architecture for distributed control of a multirobot system. In *Proceedings of the 6th International Conference on Intelligent Autonomous Systems*, pages 115–122, July 2000.
14. M. B. Dias and A. T. Stentz. A market approach to multirobot coordination. Technical Report CMU-RI-TR-01-26, Robotics Institute, Carnegie Mellon University, Pittsburgh, PA (USA), August 2001.
15. M. B. Dias and A. T. Stentz. Enhanced negotiation and opportunistic optimization for market-based multirobot coordination. Technical Report CMU-RI-TR-02-18, Robotics Institute, Carnegie Mellon University, Pittsburgh, PA (USA), August 2002.
16. M. B. Dias and A. T. Stentz. Traderbots: A market-based approach for resource, role, and task allocation in multirobot coordination. Technical Report CMU-RI-TR-03-19, Robotics Institute, Carnegie Mellon University, Pittsburgh, PA (USA), August 2003.
17. P. Doherty, G. Granlund, K. Kuchcinski, E. Sandewall, K. Nordberg, E. Skarman, and J. Wiklund. The WITAS unmanned aerial vehicle project. In *Proceedings of the 14th European Conference on Artificial Intelligence*, pages 747–755, Berlin, Germany, 2000.
18. A. Ollero et al. Control of multiple heterogeneous unmanned aerial vehicles: Architecture and perception issues in the COMETS project. *IEEE robotics and automation magazine*, 12(2):46–57, 2004.

19. E. Gat. Integrating planning and reacting in a heterogeneous asynchronous architecture for mobile robots. *SIGART Bulletin*, 2:17–74, 1991.
20. B. P. Gerkey and M. J. Matarić. Sold!: Auction methods for multirobot coordination. *IEEE Transaction on Robotics and Automation*, 18:758–768, 2002.
21. F. Giulietti, L. Pollini, and M. Innocenti. Autonomous formation flight. *Control Systems Magazine*, 20(6):34–44, 2000.
22. N. Kalra and A. Stentz. A market approach to tightly-coupled multi-robot coordination: First results. In *Proceedings of the ARL Collaborative Technologies Alliance Symposium*, May 2003.
23. E. King, M. Alighanbari, Y. Kuwata, and J. How. Coordination and control experiments on a multi-vehicle testbed. In *Proceedings of the IEEE American Control Conference*, Boston, Ma. (USA), 2004.
24. F-L. Lian and M. Richard. Real-time trajectory generation for the cooperative path planning of multi-vehicle systems. In *Proceedings of the 41st IEEE Conference on Decision and Control*, 2002.
25. M. J. Matarić and G. Sukhatme. Task-allocation and coordination of multiple robots for planetary exploration. In *Proceedings of the 10th International Conference on Advanced Robotics*, pages 61–70, August 2001.
26. M. J. Matarić, G. S. Sukhatme, and E. Ostergaard. Multi-robot task allocation in uncertain environments. *Autonomous Robots*, 2003.
27. I. Maza and A. Ollero. Multiple UAV cooperative searching operation using polygon area decomposition and efficient coverage algorithms. In *Proceedings of the 7th International Symposium on Distributed Autonomous Robotic Systems*, Toulouse, France, 2004.
28. D. Nau, T. Au, O. Ilghami, U. Kuter, W. Murdock, D. Wu, and F. Yaman. SHOP2: an HTN planning system. *Artificial Intelligence Research*, 20:379–404, 2003.
29. L. Parker. ALLIANCE: An architecture for fault-tolerant multi-robot cooperation. *IEEE Transactions on Robotics and Automation*, 14(2):220–240, 1998.
30. R.L. Raffard, C. Tomlin, and S.P. Boyd. Distributed optimization for cooperative agents: application to formation flight. In *Proceedings of the 43rd IEEE Conference on Decision and Control*, Nassau, Bahamas, 2004.
31. T. Sandholm. An implementation of the contract net protocol based on marginal cost calculations. In *Proceedings of the 11th National Conference on Artificial Intelligence*, pages 256–263, Menlo Park, CA (USA), July 1993. AAAI Press.
32. C. Schumacher and S.N. Singh. Nonlinear control of multiple UAV in close-coupled formation flight. In *Proceedings of the AIAA Guidance, Navigation and Control Conference*, Denver, Co (USA), 2000.
33. R. Simmons, T. Smith, M. Dias, D. Goldberg, D. Hershberger, A. Stentz, and R. Zlot. A layered architecture for coordination of mobile robots. multi-robot systems: From swarms to intelligent automata. In *Proceedings of the 2002 NRL Workshop on Multi-Robot Systems*. Kluwer Academic Publishers, 2002.
34. R. G. Smith. The contract net protocol: High-level communication and control in a distributed problem solver. *IEEE Transaction on Computers*, C-29(12):1104–1113, 1980.
35. J. Sousa, T. Simsek, and P. Varaiya. Task planning and execution for UAV teams. In *Proceedings of the 43rd IEEE Conference on Decision and Control*, Nassau, Bahamas, 2004.
36. A. T. Stentz and M. B. Dias. A free market architecture for coordinating multiple robots. Technical Report CMU-RI-TR-99-42, Robotics Institute, Carnegie Mellon University, Pittsburgh, PA (USA), December 1999.

37. S. Sukkarieh, A. Goktokan, J-H. Kim, E. Nettleton, J. Randle, M. Ridley, S. Wishart, and H. Durrant-Whyte. Cooperative data fusion and control amongst multiple unihabited air vehicles. In *Proceedings of the 8th International Symposium on Experimental Robotics*, Sant'Angelo d'Ischia, Italy, 2002.
38. S. Sukkarieh, E. Nettleton, J-H. Kim, M. Ridley, A. Goktogan, and H. Durrant-Whyte. The ANSER project: Multi-UAV data fusion. *International Journal on Robotics Research*, 22(7–8):505–540, 2002.
39. R. Vidal, S. Sastry, J. Kim, O. Shakernia, and D. Shim. The Berkeley aerial robot project. In *Proceedings of the IEEE/RSJ Int. Conf. on Intelligent Robots and Systems, Workshop on Aerial Robotics*, Lausanne, Switzerland, 2002.
40. R. Volpe, I. Nesnas, T. Estlin, D. Mutz, R. Petras, and H. Das. The CLARAty architecture for robotic autonomy. In *Proceedings of the 2001 IEEE Aerospace Conference*, Big Sky, Mt. (USA), 2001.
41. S. Zelinski, T.J. Koo, and S. Sastry. Hybrid system design for formations of autonomous vehicles. In *Proceedings of the 42nd IEEE Conference on Decision and Control*, 2003.
42. R. M. Zlot, A. T. Stentz, M. B. Dias, and S. Thayer. Multi-robot exploration controlled by a market economy. In *Proceedings of the IEEE International Conference on Robotics and Automation*, May 2002.

3

Communication

Volker Remuß, Marek Musial, and Günter Hommel

Technische Universität Berlin, Strasse des 17. Juni 135, 10623 Berlin (Germany)
{remuss,musial,hommel}@cs.tu-berlin.de

Summary. In this Chapter, the communication system used in the COMETS project will be presented. After an explanation of the necessities for such a system, the design idea and some technical details will be presented. The chapter ends with a look into the real system setup and a conclusion.

3.1 Introduction

When a project is about *Real-time coordination and control of multiple heterogeneous unmanned aerial vehicles* it is clear that communication between the coordinated entities whether they are on ground or flying, fixed or mobile is a crucial part.

The heterogeneous nature includes the hardware used for communication and processing. There might be different wired or wireless links and various processor architectures and operating systems. In latter case, it is a good thing to have a communication system that can work transparently on top of existing physical communication links hiding the diversity.

3.2 Requirements

Some of the key requirements that have been taken into account during the development of the Communication System (CS) are given in the following:

- Hardware:
  - The CS needed to support at least wired and wireless Ethernet and radio modems that represent a transparent serial link. To achieve a transparent system, bridging between these different links needed to be done.
  - It had to be able to interconnect standard operating systems such as Windows and Linux but also tiny 16-Bit microcontrollers without an operating system.
- Network Topology:
  - The topology of the network needed to be held flexible and dynamically changeable because the UAVs should be able to act as communication relays to each other.

A. Ollero and I. Maza (Eds.): Mult. Hetero. Unmanned Aerial Vehi., STAR 37, pp. 49–65, 2007.
springerlink.com

  - For robustness, the CS should be able to build independently working subnetworks in case of a network split when for example a group of UAVs is flying out of communication reach. These subnetworks should be able to merge again later.
- Functionality:
  - The CS needed to transport data reliable from one sender to one or more receivers.
  - Bandwidth constraints on an information flow level needed to be fulfilled including maximum delays for real-time communication for single network hops.
  - The transportation should happen fully transparent under all circumstances of changing topology and physical links as long as there is an existing path between the sender and the receiver.
  - Only when sender and receiver are unreachable to each other caused by a network split, sender and receiver needed to be informed.

3.3 Design Idea

To fulfill the network topology requirements it was necessary to design a CS that is able to work without a dedicated master or any other central entity in the network. Due to this and the ad-hoc nature of changing UAV configurations the decision was taken to implement a full multi-hop ad-hoc network scheme that supports dynamic routing.

The term *user* in the following refers to the programmer using the communication system's application programming interface (API).

3.3.1 Data Organization

To transmit data this network makes use of a distributed shared memory (DSM) approach and is called blackboard communication system (BBCS). The blackboard BB is a data structure virtually shared between all network nodes using its coherency protocol (CP). The CP makes use of a platform abstraction layer (PAL) that adapts the CP to different architectures, operating systems and network layers. The system is able to overlay on any byte transmitting link, so the design includes also transport layer functionality. A partly similar design has been identified [4].

3.3.2 Network

A network node (NN) is an entity that is taking part in the communication. A NN is a software task that can run on a multi-tasking operating system (OS), a single tasking OS or just a routine on a system without any OS.

A port is a point-to-point connection between two NNs and every NN can act as a relay. In this way, a multi-hop ad-hoc topology of the network can be realized [7]. This design basically goes back to the DARPA packet radio networks

[3]. Every port can consist of a flexible number of channels where each channel is a unique bidirectional byte transmitting link.

Channels have a restricted bandwidth (CBW) assigned which has to be the minimum bandwidth of the underlying physical layer to achieve guaranteed bandwidth communication. A port's traffic is distributed among its channels, so that the bandwidth of a port is the combined bandwidth of all its channels. More than one channel per port will also enable connection redundancy and therefore strengthen robustness of the connection between two NNs. A port is treated as active as long as it has a working channel; otherwise inactive.

Channel activity is based on traffic and timeouts. A NN sends either data or a dedicated keep-alive message through an active channel. If there is no incoming traffic for a fixed but variable amount of time, the channel is regarded as inactive until new traffic is detected.

3.3.3 Blackboard

All NNs have access to the BB. It consists of so-called slots where each slot is associated with a specific class of information and can hold a specific amount of data. As with variables in conventional memory, mutual exclusion accessing slots is necessary. A slot can have an infinite number of data drains (aka subscribers) but only a single data source (aka publisher) on the network. Each slot has a unique id and an assigned relative slot bandwidth (SBW), which is given as a fraction of the CBW. Assigning different SBWs allows emphasizing vital data.

The structure of the complete BB is fixed during execution time and in a real system usually defined during compilation time. Nevertheless, it is not required that every NN has a local copy of the complete BB. Only slots that the node is publishing, subscribed to or relaying (see below) are held locally. It is clear that the summed up bandwidth of all slots must not be more than 100 % for every slot of an NN.

The approach is somewhat similar to a topic-base publisher/subscriber system, but is completely free of any central entity [2].

3.3.4 Routing

The routing in the network is performed individually for every single slot from the source to all drains. It makes use of a visibility graph, which is distributed in the network nodes. The metrics usually used is the number of hops. The approach is known as distance vector routing [5]. Every NN in the BBCS denotes its distance of the source for each slot. This information is available to all adjacent nodes. If a NN is starting to read a slot, it becomes a data drain. The BBCS then requests the data at the adjacent NN with the minimum metric (hop count) for this slot. This neighboring NN then follows the same procedure and the route is built until the source NN is reached.

The result is a distribution tree for every slot where the root is the data source for this slot, all leaves are data drains. Any other node is either a data drain or a relaying NN.

In the case of a network topology change due to a state change of a port from active to inactive or vice versa, the new routing information will be distributed through the network and may reconfigure the path.

For every NN the user can decide on slot base if it will provide relaying service or not for this particular slot, whereas relaying itself is transparent for the user.

The BBCS uses message replication while routing (MRWR) approach for multicasting [1]. Even if there are several data sinks for a single data source in a network, data will only be transmitted once between hops as long as the routing trees are matching. Data multiplication for n data sinks will always occur in the node that is closest to the sink and present in each of the n routing trees. As

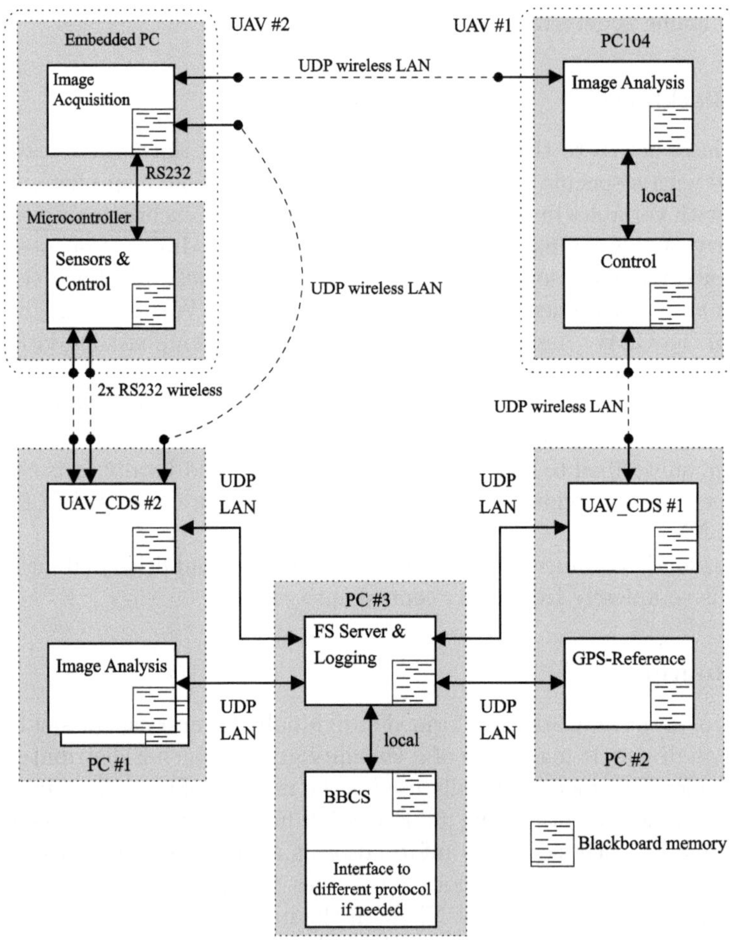

Fig. 3.1. Sample configuration of the BBCS. Network nodes are shown as white boxes, computers are shown as grey boxes, the communication links are represented by arrows, and the squares with horizontal lines depict the local copies of the BB part in each of the NNs.

a result, the same data is only transferred once on a single channel which is a necessity for the BBCS's bandwidth management approach.

3.3.5 Network Layers

The CP copes with flow control, out-of-order reception, lost packages, error and failure detection. Therefore it implements OSI layers 6 to 2 [8]. There is only the need for an additional physical layer. A physical layer could be virtually everything capable of transferring bytes of data, for example a RS232 serial link. Nevertheless, during the system design it became clear that it is a more efficient approach in many cases to overlay on a higher layer level to use existing infrastructure. The decision was taken to implement the CP also on top of existing layers such as the transportation layer.

In the system used during the COMETS project, the CP was used on bare serial wired and wireless RS232 links where it incorporates layers 6 to 2 as well as over wired and wireless UDP and TCP connections where it resembles the network layers 6 to 4. The benefit of using existing layers 3 to 1 is the opportunity to use all standard hardware and the possibility to test the communication even over long distance internet links.

3.3.6 Network Setup

In Fig. 3.1 an example setup is shown that corresponds to a possible setup in the COMETS project as it was anticipated during the design phase of the CS.

Using socket-based communication as underlying network, the NN processes of the base stations may be distributed among any number of computers or run on a single PC. The software has not to be changed when the physical distribution changes. It is possible to start a new NN at any time with a parameter indicating to which of the computers it should connect. It is equally easy to connect the UAV simulator instead of the real UAV to the system. Thus hardware-in-the-loop simulation can easily be performed.

3.4 Design Details

This section will feature some key methods and algorithms used to design the blackboard communication system with respect to practical issues.

3.4.1 User View

The user accesses the slots in the BB by using the API which comes for C. There are two different types of slots:

State-type information slots: are record expressions that represent atomic data. These slots are directly mapped to variables accessed by the user program. In practice, these variables are mostly records that contain some kind of system state, e.g. a set of telemetric data, sensor measurement or just the current time. Although this is application-dependent, the typical state information consists of a single self-contained data record, relatively small or at least of limited size. Very often, only the latest data record available is of relevance, since earlier records are outdated. It may also be unnecessary to propagate every change to the data record successfully through the network as long as some user-defined upper bound on the transmission delay is met.

Stream-type information slots: are basically consisting of a sequence of bytes. These slots are not directly accessible by the user and API calls have to be used to access these. Slots of this kind are used to transmit any kind of digital stream, for example digital images, video streams or streams of DGPS corrections. The typical characteristic of stream-type information is that they are not semantically self-contained data units. They are relatively large and varying in size and they have to be transmitted uninterrupted and uncorrupted for usability.

The decision to support these two types of information semantics was taken because hybrid forms are extremely rare, at least in the field of UAV control. The API does provide these types of service, one for state-type semantics and one for stream-type semantics.

Since state type slots are directly mapped to variables, accessing them is straightforward for the user. Another advantage is that the slot has the same definition of the state type slot contents at every NN and there is no need to parse streams to gather the data. The BBCS is automatically filling the variables. In addition, one has to be aware of differences in data structure sizes imposed by the use of different architectures and compilers beforehand.

Stream type slots are accessed by API calls that are similar to usual file access calls for reading and writing. The BBCS maps these data stream accesses internally into fixed-size slots.

Insecure and secure transmission schemes are supported by the BBCS. A secure transmission means that full end-to-end, error free transmission is guaranteed by the BBCS. Insecure transmission means that no end-to-end control is performed. In this scheme, data can be lost on the way or being overwritten by succeeding data in the same slot. Insecure transmission still keeps the contents of a slot consistent. It shows that in practice, data in state type slots can often be transported as insecure because for example, only the latest set of telemetric data is of interest and there is no sense in delaying the most current telemetric data in order to retransmit an older one. Data transported in stream type slots is in most cases data that has to be transported securely because the stream has to be uncorrupted to be of value, e.g. transmission of images. However, the exception proves the rule. A state type information that should be transferred as secure is e.g. a command and stream type information that should be transferred as insecure are for example DGPS corrections.

Mutual exclusion is necessary when the BB is accessed by the user and the CS. The design decision was taken to have a special API call that triggers the synchronization and propagation over the network. This approach has been chosen because it is important to guarantee data integrity for each slot. Updating a number of components of the same slot has to form an atomic transaction. Therefore, the explicit invocation of the service for updating is actually the most convenient and least error-prone solution. The programmer does not need to care about concurrency, because the use of the BB variables between two invocations of a synchronization operation is always atomic.

3.4.2 Conditioning of Slot Contents

This section outlines a few key aspects of the CP. The data exchange perspective as presented here refers to the data stream belonging to a single BB slot and being sent over a single BB port. Multiplexing of the data streams of multiple slots will be described in 3.4.3.

A hierarchy of three levels can be identified in the CP data stream sent for the synchronization of one slot. They are, listed in low-to-high order:

Stripes: are single uninterruptible data packages of a fixed maximum size. The maximum stripe size will typically range from 64 to 512 bytes and is a global BBCS parameter. Stripes are in no way visible to the user. Error detection code is used to assure the integrity of stripes. A sequence-numbering scheme is used to identify stripes and permit the reordering after out-of-order reception of stripes.

Units: are groups of stripes; the maximum number is a global configuration option of a BB slot. A unit contains the atomic contents of a slot. When using state-type transmission, a unit is the amount of data exchanged in an atomic transaction.

Messages: may be composed of one or more units. They are represented by setting a flag on the last stripe of the message, called the End-of-Message (EOM) marker. A stripe with EOM set always terminates the current unit. When using state-type transmission, the EOM is set at every unit posted, while during stream-type transmission the user has control of the EOM markers. The use of the unit structure is actually redundant for the stream-type service, but results from the fact that the BBCS implementation uses a platform abstraction layer and a common internal layer for unit stream transmissions, while the service type is only distinguished within a small interface layer on top thereof. If BBCS transmissions are relayed via intermediate NNs that do not access these slots, only the service-independent layers are involved. Figure 3.2 visualizes this situation for a transmission from NN#1 to NN#3 using NN#2 as relay.

3.4.3 Data Transmission

This section deals with the actual data transmission, the approach and capabilities of the BBCS in regard to real-time requirements and bandwidth control.

The data transmitted over each channel has to be a statistical multiplexed stream that consists of the slot data. The packet size equals the stripe size as

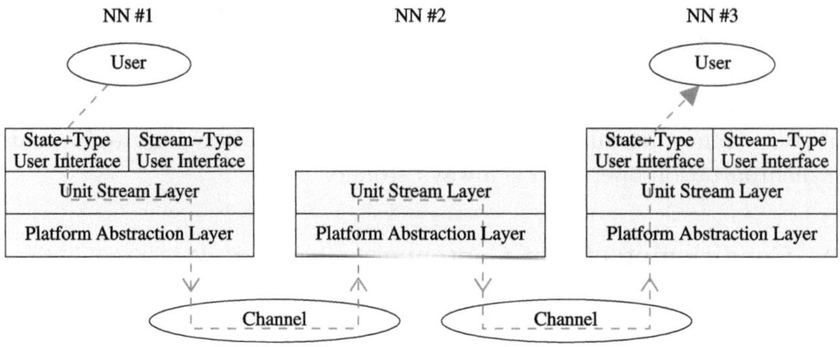

Fig. 3.2. BBCS Layers involved in transmission over intermediate NN

mentioned above. The multiplexing schedule needs a packet scheduling algorithm that takes into account the bandwidth fraction attached to every slot. Using communication media with a non-zero error-rate like wireless connections as physical layer does not allow guaranteeing hard real-time constraints for message transmissions. Still, the communication process itself has to be considered reliable here, since real-time aware physical layers are available as well. The queuing delay, the time between the production of a data record and its delivery to the communication channel, has to be analyzed for their worst-case timing.

Whenever a channel's assigned bandwidth permits to send the next stripe of data, the BBCS has to schedule the slot that shall be allowed to send. This multiplex is calculated using an approach to static proportionally fair scheduling named static binary tree bandwidth scheduling (SBTBS). Due to the slot's static bandwidth assignments, the schedule can be semi-static and SBTBS is a very efficient way to calculate it.

The user specifies the share of CBW to be reserved for every BB slot. This share is given in multiples of a fraction of f, where f is the resolution of the schedule and needs to be a power of 2. The sum of all slot bandwidths has to be less or equal to 100%. Figure 3.3 gives an example of the bandwidth reservation using $f = 16$ and the given bandwidth assignments for four slots in the table on the right. From these values, the SBTBS algorithm calculates the binary tree in the figure.

Its leaves are marked with slot numbers, its inner nodes with the denominators of the fraction of bandwidth they represent (a node marked 8 represents 1/8 of the bandwidth available). The length of the path to every leaf also indicates the bandwidth occupied by the leaf in question. The tree yields a sequence of $f = 16$ slot numbers by selecting paths to leaves according to the binary representations of the sequence position numbers $0 - 15$, least significant bit first. The resulting static transmission schedule is applied repetitively. It meets the essential criteria: The schedule fulfils the bandwidth requirements for all slots. All slot numbers are well distributed over the schedule, resulting in minimum worst-case queuing delays.

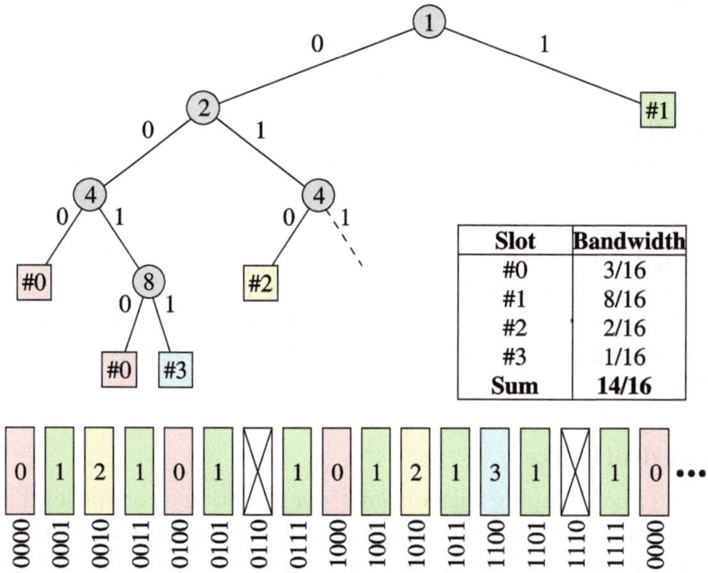

Slot	Bandwidth
#0	3/16
#1	8/16
#2	2/16
#3	1/16
Sum	**14/16**

Fig. 3.3. Static Binary Tree Bandwidth Scheduling (SBTBS) of stripe transmissions

The CBW is never exceeded while processing the schedule. The result is traffic shaped, burst-free data output at the CBW similar that is leaky bucket constrained [6]. As one can see, for evenly distributed SBWs the result would be a round-robin scheduling.

Note that bandwidth assignments where the numerator is not a power of 2, like $3/16$ for slot #0 in the example, result in the reservation of multiple nodes in the scheduling tree. Taking this into account, the upper bound on the queue delay d_S between two subsequent transmissions for a slot S with a reserved bandwidth of r/f can be identified, in multiples of the stripe transmission time, as:

$$d_S < \frac{2f}{r} \tag{3.1}$$

If, however, the numerator r is a power of 2, one gets

$$d_S \leq \frac{2f}{r} \tag{3.2}$$

as an upper bound, which is obviously the optimal upper bound achievable.

In practice, relevant queuing delays may also be influenced by the size of a data unit, which may consist of several stripes, and the number of intermediate NNs on the route between data source and drain. In this more general case, the source NN might have to finish the currently active transmission of a unit of slot S before starting the transmission of a newly generated data unit. For units of k stripes travelling n hops on the route, with the same physical bandwidth and the same bandwidth share r/f assigned at each NN, the total queuing delay D_S

until complete reception, in multiples of the stripe transmission time, is limited as follows:

$$D_S < (2(k-1)+n)\frac{2f}{r} \tag{3.3}$$

This includes 3.1 as a special case. It can be proven that SBTBS as outlined above works for every bandwidth distribution not exceeding 100% in total, and that the worst-case run-time complexity of the SBTBS scheduling algorithm is $\mathcal{O}(f)$. Free places in the static schedule, as at positions 12 and 14 in Fig. 3.3 can be safely skipped at run-time in favor of the next entry of the schedule, which does not affect the time bounds in any negative way.

In a real system such as in the COMETS project a good choice for the schedule resolution is 1024 since then the SBW can be conveniently specified in thousandth.

3.4.4 Physical Links

As mentioned above, coping with out-of-order reception on stripe level and doing cyclic redundancy checking, everything capable of transmitting streams of bytes is able to act as an underlying network connection or link. The CBW has to be chosen to match the links capabilities. Ideally the link will be able to sustain the defined CBW at all time. For wired links such as wired Ethernet using UDP or TCP it is easy to come up with a CBW that can be guaranteed as long as the only ongoing communication is done using the BBCS, which is a feasible assumption in dedicated systems such as a multi-UAV ground station. Also radio modems (as used in the COMETS project) that act as transparent point-to-point connection have proven to give an all or nothing type of connection.

Usual wireless Ethernet links on the other hand are of self-adapting speed and to choose a CBW that can be supported at all time will result in a very low speed or is even impossible.

To cope with these weak links, the BBCS can make use of graceful degradation for every channel on user request. If enabled, the NNs are allowed to reduce the bandwidth assigned to a channel. The decision will be made based on statistical information about the percentage of stripe loss that will be generated by the receiving NN at the other end of the channel and reported to the sending NN. The user will always stay informed about the currently assigned bandwidth. Fall-forward can be done via timeout or triggered by the user, which is a good solution because the user might have better knowledge of the underlying physical layer, e.g. the link quality of a wireless LAN.

3.5 Implementation Details

In the following some implementation details will be pointed out.

3.5.1 Core Architecture

The Black-Board-Communication-System (BBCS) is implemented as a central core that takes the users API calls as input and outputs n traffic shaped bytestreams for n channels. These bytestreams are transmitted using a platform abstraction layer (PAL) that interfaces the central core to existing transportation or physical layers. The BBCS core is needed once in each NN in a running system.

3.5.2 Cross Platform Support

The underlying protocol used to assure the integrity of the transported bytestreams makes use of cyclic redundancy checking (CRC). The protocol itself uses network byte order and the BBCS copes with the byte order of the machine running the NN using the PAL.

Nevertheless, as always, when it comes to the data transported, the user still needs to handle byte-order differences in user structures. Conversion of structures of different byte-orders would be possible when the user attaches a descriptor to every state type slot that denominates the logical structure of the slot's content. Automatic generation of these descriptors is only possible during compilation time and therefore needs support of the compilers used. It has not been evaluated if such a compiler already exists since the decision was made to use the standard compilers. Conversion of stream type slots is not an issue because they are defined as a bytestream transportation slot.

3.5.3 Support for Embedded Systems

To support not only full-blown OS but also single task systems like microcontrollers, the implementation is single-threaded and based on 16-Bit integers algorithms. The usage of static and dynamic memory is very low since there is done very little internal buffering in the BBCS. The dynamic memory is allocated for control structures and only done during the setup phase. Allocation is done using the PAL. On systems where these calls cannot be mapped to OS calls, usually because there is no OS, the PAL has to implement a memory allocation scheme. Since memory is only allocated but never freed, the implementation is simple.

The Black-Board (BB) itself is composed from user data and its content is directly copied into the output buffers of the according transportation or physical layers. Since these buffers are part of the PAL, their size can be chosen to match the systems capabilities. The size of a single stripe, which is 256 Bytes in COMETS, is the minimal size for the output buffers.

3.5.4 Implemented Physical Abstraction Layers

During the project the PAL has been implemented for three different systems and three different transportation/physical layers.

For use as transportation layers, TCP and UDP are implemented. Using these protocols makes it possible to use existing infrastructure such as the Internet and existing network hardware such as wired and wireless Ethernet. In addition, the Serial ports (RS232) are supported as physical layer.

The PAL implementation including TCP, UDP and Serial is available for Windows (win32 API) and Unix (POSIX API). They have been tested on Windows XP, Windows XP embedded, Windows 2000, several i86-Linux, Mac OS X and SUN Solaris.

The PAL implementation for the microcontroller Infineon C167 only supports Serial as physical layer.

3.5.5 Memory Efficiency

The BBCS is quite small. For example the footprint of the C167 version is about 30 kBytes. This makes the BBCS feasible for use in a lot of embedded systems and even for the so-called smartdust systems.

3.5.6 Code Example

Figure 3.4 shows a piece of pseudo-code to depict the state-type handling of BB slots. It contains a very simple but still complete proportional (P) controller controlling the throttle servo position of an UAV in order to keep a desired altitude `A0`. During initialization, the program specifies that the state variable `bb.gps` shall be permanently updated from the network and the state variable `bb.servo` shall be posted to all NNs needing it whenever possible. Later, the program enters an endless loop where it first reads the current altitude from the BB by just accessing the variable `bb.gps`. After calculating the resulting throttle output `thr`, that value is assigned to the corresponding component of the BB state variable `bb.servo`. At last, the synchronization of the BB states is initiated using the `bb_sync` operation.

```
bb_auto_get(BB_gps);                // init to receive bb.gps...
bb_auto_put(BB_servo);              // ...and send bb.servo automatically...
LOOP                                // endless main loop:
  alt          := bb.gps.z;         // get altitude from blackboard
  err          := alt - A0;         // calculate error of altitude
  thr          := T0 - P*err;       // calculate P controller answer
  bb.servo.thr := thr;              // write answer to blackboard
  bb_sync();                        // trigger reception and propagation
END LOOP;
```

Fig. 3.4. User perspective – Possible user code for the handling of state data

Figure 3.5 shows a piece of pseudo-code to depict the stream-type handling of a BB slot. It is a fully functional loop to transmit, for example, a stream of DGPS reference data received on a serial port to a UAV, which requires this stream to obtain more accurate position information from the on-board GPS receiver. The byte array `buf` is used to hold data read from the serial port but not yet sent via the BB. The variable `len` always contains the number of bytes present in this buffer. The `bb_write` operation, which is the focus of this example, returns the number of bytes that could be written to the `BB_gpsref` slot of the BB.

In both code examples above the user view for the usage is shown. The similar simple setup code is omitted.

```
len    := 0;                                   // initially: buf is empty
LOOP                                           // endless main loop:
  len += read(f, buf+len, sizeof(buf)-len);    // append data from file
                                               // descriptor to buf
  out := bb_write(BB_gpsref, buf, len);        // send as much data
                                               // as possible via BB
  len -= out;                                  // which makes buf more empty
  memmove(buf, buf+out, len);                  // move remaining data
                                               // to start of buf
  bb_sync();                                   // trigger propagation
END LOOP;
```

Fig. 3.5. User perspective – Possible user code for the handling of stream data

3.6 Real Scenario

During the COMETS project field experiments, the different partners (LAAS, GMV, HELIV, AICIA, TUB) had to communicate using a local communication setup. In the following, a description of the real COMETS setup will be given including network topology, transportation means, and information flow.

3.6.1 Network Topology in COMETS

In the BBCS, a NN has the possibility to relay communication without user intervention on condition that the slot to be relayed has been made known to the node. In COMETS this functionality was used to ease-up the network topology and the bandwidth distribution between the communicating partners.

The network relay node (NRN) was introduced. The NRN is designed in a way that makes it possible to use a common source code for every partner. To achieve this, the common NRN had to be aware of all BBCS slots that are used between any partners. Designed that way, the NRN is able to relay all inter-partner communication done in the COMETS network. Furthermore every partner can easily access information that has already been added by another partner without the need to do a bilateral setup.

Since every partner is hiding all its internal nodes behind the Network Relay Node (NRN) - as shown i.e. in Fig. 3.6 for partner TUB - there only have to be defined 10 channels at maximum for a fully interconnected network with 5 partners (see Fig. 3.6).

The setup of the internal nodes can be done by any partner individually as long as every partner uses its allocated sector of the unique slot id space. These internal networks are also including links to UAVs.

The topology of the NRN network can be matched on a fully switched Ethernet setup (see Fig. 3.7). Since all ongoing communication is using the traffic shaped connections with the NRN, real-time constraints can be guaranteed as long as a sensible bandwidth had been chosen. During the experiments the biggest amount of data transported was image data for analysis which needed 400 kB/s. Therefore, it was sufficient to use a bandwidth of 600 kB/s between the NRNs. The accumulated bandwidth of 4800 kB/s per node is easy to handle for

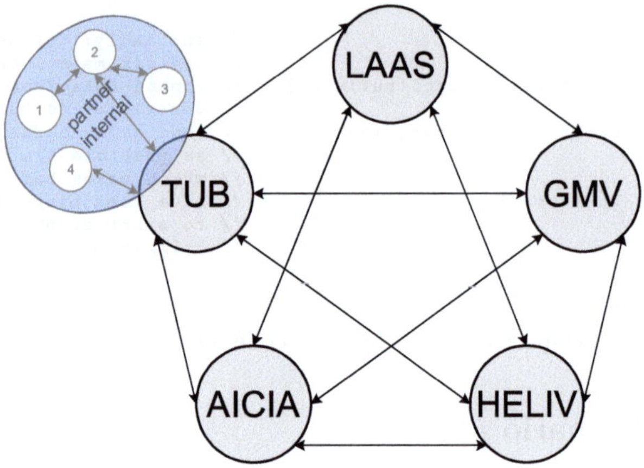

Fig. 3.6. COMETS BBCS network topology using a network relay node

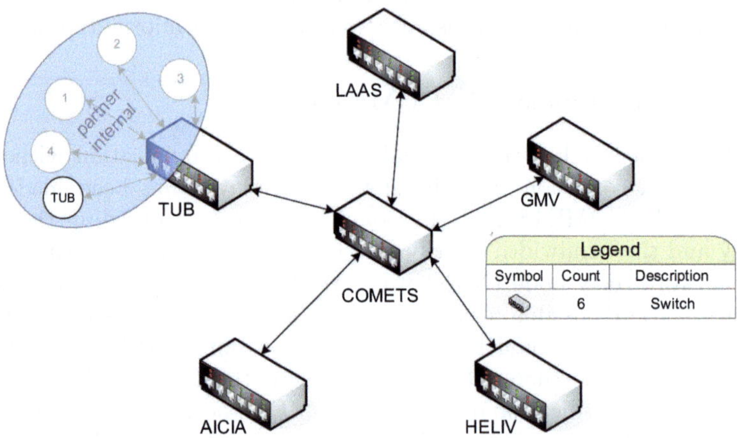

Fig. 3.7. COMETS Ethernet network topology during field experiments

the fully switched 100 Mbit/s Ethernet setup because due to the traffic shaping and switching, the Ethernet communication flow in the NRN network will be practically collision free.

3.6.2 Transportation and Physical Layers

In the beginning of the development, TCP was used to establish all Ethernet links. Later, it turned out that using UDP is a more meaningful solution for COMETS and the BBCS. The use of TCP has several drawbacks:

- TCP has server/client architecture which makes it necessary to decide if the node is a TCP server or client. Since in the BBCS every node is equal, this is an useless distinction.

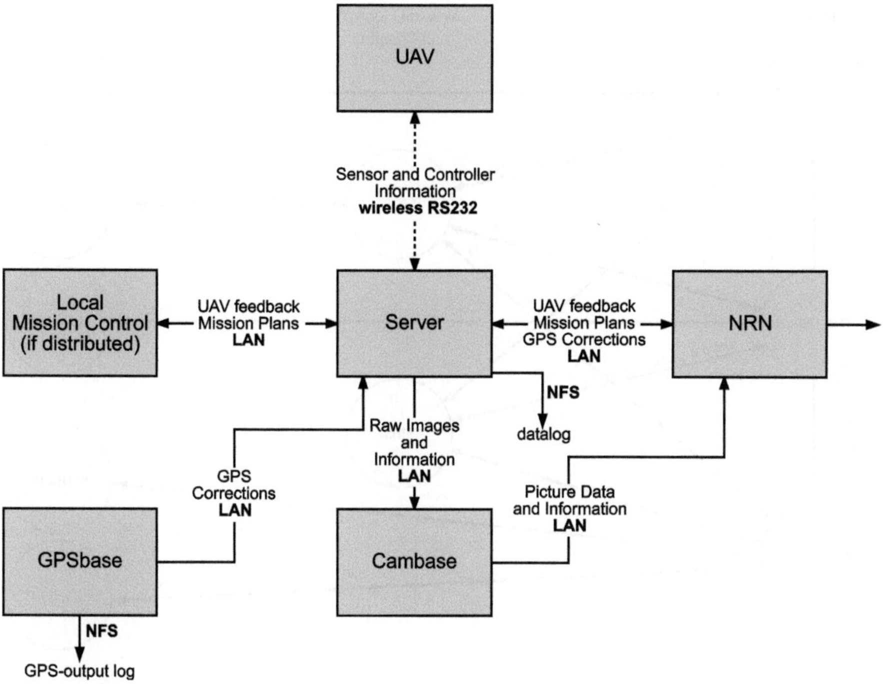

Fig. 3.8. Example of a partner internal BBCS network behind the NRN

- When the connection to a link partner is lost, TCP requires user code action on both sides of the connection to reestablish the link. That contradicts the idea of transparent, automatic routing.
- TCP is a reliable protocol that uses error checking and flow-control. In combination with the BBCS this is a hitch. The BBCS can manage all this on his own. Additionally TCP interferes with the real-time idea of insecure slots by delaying new data because of old data that had a transmission error.
- TCP is known to perform bad over wireless Ethernet. The flow-control scheme usually implemented in common TCP stacks is unable to adapt efficiently for this kind of connection.

On the other hand, UDP is unreliable, has no flow-control and it can be configured to ignore the absence of a link partner, so it is a perfect match for the BBCS. The decision was taken to use only UDP as connection type over Ethernet in the COMETS project. Serial links are used in the COMETS project as well.

3.6.3 Information Flow

In Fig. 3.8 a typical information flow for an existing partner internal network is shown. Every block is a single BBCS NN and there is a lot of communication going on in the network using reserved slot ids. Some local data logging is done and communication to the rest of the system using the NRN.

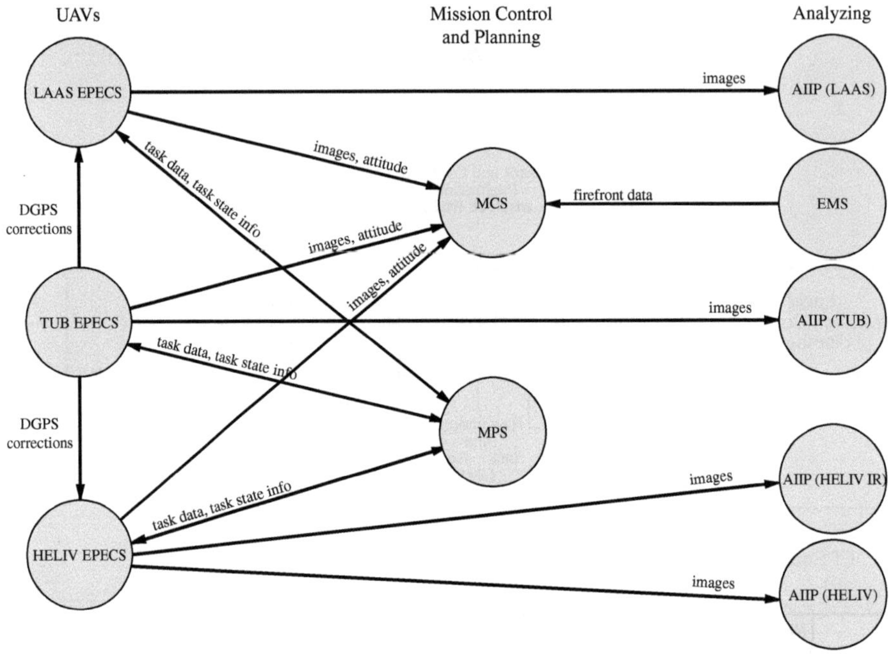

Fig. 3.9. Information being sent in the network between partners and NRNs

Figure 3.9 depicts a typical information flow situation in the COMETS project during mission execution. Only information flows are shown that are using the NRN network to be transmitted between partners. Every circle is representing an information source or drain in the BBCS network but is not necessarily exactly one NN.

Left hand side, the UAV representatives are shown, in the middle there is the centralized mission control and planning for the full system and on the right, the analyzing of gathered data and its feedback.

3.7 Conclusion

The communication system designed, implemented and tested during the COMETS project incorporates several known design ideas and some new. In difference to the mostly scientific work done in this area it is always designed with the practical view in mind and includes everything needed for a working system from the highest to the lowest level. It is a complete solution for many kinds of communication needs.

The BBCS is a good solution for single serial links between microcontrollers as well as inter-process communication and workstation applications. It has proven to work reliable and useful in all these scenarios at the same time while being used in mission critical functions in a fleet of UAVs.

References

1. S. Bhattacharya, G. Elsesser, Wei-Tek Tsai, and Ding-Zhu Du. Multicasting in generalized multistage interconnection networks. *J. Parallel Distrib. Comput.*, 22(1):80–95, 1994.
2. P. T. Eugster, P. A. Felber, R. Guerraoui, and A. Kermarrec. The many faces of publish/subscribe. *ACM Computing Surveys*, 35:114–131, 7 2003.
3. J. Jubin and J.D. Tornow. The DARPA packet radio network protocols. *Proceedings of the IEEE*, 75(1):21–32, 1 1987.
4. F. Peña-Mora, R. Aldunate, and M. Nussbaum. Availability analysis of an ad-hoc DSMS for disaster relief environments. In *SCCC '02: Proceedings of the XII International Conference of the Chilean Computer Science Society (SCCC'02)*, page 59, Washington, DC, USA, 2002. IEEE Computer Society.
5. Andrew S. Tanenbaum. *Computer Networks, Third Edition.* Prentice Hall, Upper Saddle River, New Jersey, 1996.
6. J. S. Turner. New directions in communications (or which way to the information age?). *IEEE Communications Magazine*, 40(5):50–57, May 2002.
7. D. Tutsch. *Performance Analysis of Network Architectures*, page 55. Springer, 2006.
8. H. Zimmermann. OSI reference model - the ISO model of architecture for open systems interconnection. *IEEE Transactions on Communications*, 28:425–432, 4 1980.

4

Multi-UAV Cooperative Perception Techniques

Luis Merino[1], Fernando Caballero[2], Joaquín Ferruz[2], Johan Wiklund[3], Per-Erik Forssén[3], and Aníbal Ollero[2]

[1] Pablo de Olavide University, Crta. Utrera km. 1, 41013 Seville (Spain)
`lmercab@upo.es`
[2] Robotics, Vision and Control Group, University of Seville, Camino de los Descubrimientos s/n, 41092 Seville (Spain)
`{caba,ferruz,aollero}@cartuja.us.es`
[3] Linköping University, SE-581 83, Linköping (Sweden)
`{jowi,perfo}@isy.liu.se`

Summary. This Chapter is devoted to the cooperation of multiple UAVs for environment perception. First, probabilistic methods for multi-UAV cooperative perception are analyzed. Then, the problem of multi-UAV detection, localization and tracking is described, and local image processing techniques are presented. Then, the Chapter shows two approaches based on the Information Filter and on evidence grid representations.

4.1 Introduction

Applications such as natural and human made disasters scenarios, search and rescue, law enforcement, aerial mapping, traffic surveillance, inspection or cinematography require robust and flexible perception systems [31]. These perception systems use the sensors on-board the UAVs to perceive the environment, estimate the situation of an event and/or the own state of the robots.

In the case of the team of robots, one can do better than a robot perceiving alone. The information that each robot obtains about the environment can be *shared* to improve the perception, so that each robot obtains a better picture of the world than if it would be alone. Moreover, from the perception point of view, the robots can explicitly *cooperate*, developing actions to collect data. Thus, *cooperative robot perception* could be defined as the collaboration inside a fleet of robots for the estimation of the state of the environment, by sharing information or even by developing cooperative actions.

In this chapter we consider multiple heterogeneous UAVs in cooperative perception activities (Fig. 4.1 illustrates a possible scenario). The heterogeneity increases the complexity of the problem, but also provides several advantages for the application such us the possibility to exploit the complementarities of different UAV platforms with different mobility attributes and also different sensor and perception functionalities. It should be noted that many applications require several sensors that cannot be carried by only one UAV due to payload limitations.

A. Ollero and I. Maza (Eds.): Mult. Hetero. Unmanned Aerial Vehi., STAR 37, pp. 67–110, 2007.
springerlink.com

Fig. 4.1. Different moments during a fire detection mission using three UAVs

In these cases the cooperation between the UAVs, equipped with different sensors, should be established also at a perception level.

4.1.1 Main Issues

When considering multi-robot perception, there are several important issues that should be addressed.

Knowledge representation and information fusion. Robots will use their sensors to update their knowledge. This knowledge is usually arranged into a hierarchy from lower to higher abstraction levels, ranging, for instance, from the raw sensor readings, estimations of the own robot position and velocity or the position of the surrounding obstacles to the estimation of the shape or appearance aspects of an object or the identity of a particular object.

If the robots can communicate, then the information received from other robots can be fused with the local one to improve this knowledge. Information fusion is, then, a key issue in a cooperative perception system.

Information fusion requires to translate the received information to the same local representation. Therefore, the general fusion architecture will affect the local representation employed by each robot. Also, related to that is the problem of data association, that should be solved in order to determine if two robots are referring to the same part of the world.

Fusion rules should lead to an improved knowledge about the world, but care should be taken to avoid *rumor propagation* in the case of decentralized systems. This can occur within a fleet of robots, and can lead to overoptimistic estimations.

Cooperation. Robots use locally the perception estimates to react under a changing world and even to develop plans if they are endowed with decisional abilities. These actions can include information gathering tasks that improve the local knowledge of the world, like visiting unknown zones or moving to better points of view. Moreover, the robots can coordinate themselves and even cooperate in these tasks. In this case, it should be considered *metrics* about the gain of information that the actions of a particular robot of the fleet produce from the point of view of perception.

4.1.2 Approach and Outline

The framework that will be used for knowledge representation is probability theory. It is accepted that any system capable of uncertain reasoning is generally more robust that other one that does not [46]. So, the perception algorithms should be able to handle uncertain data, and to maintain information about this uncertainty. In the last decade, probabilistic algorithms have become one of the most prominent tools for uncertain reasoning among roboticists. In probabilistic perception, the current knowledge is represented by a probability distribution that assigns to each possible value of the state space a probability, the *belief state*.

Regarding the multi-robot perception architecture, the first decision is whether the information gathered by the fleet is combined in a purely centralized fashion (all the data collected in a central node that builds a world representation) or in a decentralized manner, in which each robot compounds its own model and communication occurs at a higher level. Also, the architecture considered could be something in between these two extremes. Of course, the solution depends on several issues, like the physical communication layer available, the local processing power of the robots, the tasks to be accomplished, the structure of the environment, the autonomy with which the individual robots should be endowed, etc.

In general, centralized solutions can provide optimal solutions, but they do not scale well with the number of robots. Therefore, the proposed multi-robot perception architecture is decentralized, in which each robot builds its own local (and possibly partial) representation of the world, which is described by the belief state. Then, the robots will share their beliefs, and include the beliefs received from other robots in order to improve their knowledge. No central node is required, although the information is transmitted to a central station for planning and replanning activities, operator visualization, etc.

The rest of the Chapter is organized as follows. The probabilistic framework for decentralized perception is summarized. This framework will be applied for cooperative detection, localization and tracking. The main sensors considered are video cameras. Then, the local processing techniques employed by the UAVs are described. A first approach, based on the information filter, is presented. Finally, a grid-based approach for fusion of heterogeneous sensors is described.

The perception system (see [25] for a more detailed description of the software system), together with the decisional architecture presented in Chap. 2, allows for cooperative detection, confirmation and localization of events of interest. Chapter 8 will present the application to forest fire detection. This is a very relevant application in many countries where forest fires have disastrous social, economic and environmental impact.

4.1.3 Related Work

An important part of the work on multi-robot systems has been focused on the development of architectures for robot coordination. Several architectures have been summarized in Chap. 2. In general, these architectures focus on task planning, task allocation, coordination, planning and control, conflict resolution, etc. Although a group of robots can coordinate with no or little communication [24, 4], all in all the previous architectures require communication of actions, plans, tasks, which can be interpreted as a kind of information sharing. However, the knowledge sharing problem and its implications on robot coordination are not explicitly considered in most of those approaches.

Of course, there have been applications of robot teams for distributed and/or cooperative perception activities. For instance, in the task of map making [39, 38, 15, 5]. More recently, approaches including cooperative multi-robot Concurrent Mapping and Localization (CML) have been presented [11, 47].

Other applications include multi-robot surveillance. For instance, the objective of the CyberScout project [35] is the development of a network of robots for reconnaissance and surveillance operations. Multi-robot target tracking is considered in several multi-robot systems, like [33, 44]. More recently, Howard [19] has described results within the DARPA Software for Distributed Robotics initiative with an heterogeneous team of near 80 robots in indoor experiments. Mapping activities, SLAM and cooperative detection of intruders are applications considered. In the context of Robocup several cooperative perception results have also been presented, as for instance [43, 52, 36].

Many of the applications are ad-hoc algorithms for combining data from several sources. The different approaches differ in the way they represent the information, how the data is communicated and fused, and the network topology (centralized or distributed). References [28, 33] deal with issues of decentralized information fusion employing probabilistic techniques.

Most previous approaches are applied in structured environments. Regarding unstructured environments, less work can be identified. Moreover, the application of cooperative perception for teams of UAVs are more rare than in teams of ground robots.

In the BEAR project [50], pursuit-evasion games involving UAVs and Unmanned Ground Vehicles (UGVs) are considered. A probabilistic framework is employed for obstacle and evaders position estimation by the fleet of pursuers (UAVs), and to determine pursuit polices [51]. In this case, the data fusion and policy determination are carried out by a central node.

Closest to part of the work presented here is the work described in [45] developed in the framework of the ANSER project. In that project, an architecture for multi-vehicle data fusion is designed, and its application to multi-UAV SLAM using vision is presented. State estimation is performed distributelly using the information form of the Kalman filter. Each vehicle uses the information received to update its state and its local map. The map information (represented as the location of a discrete set of landmarks) estimated for each vehicle is propagated to the rest of the fleet. In this work, artificial landmarks of known size are used in order to obtain range and bearing measurements. Nevertheless, the main issues regarding decentralized data fusion are considered.

In the context of the same group, Reference [17] presents techniques for coordination of sensor platforms in order to maximize the information gain, which is important for cooperative perception activities. It describes results derived from the information form of the Kalman filter or Information Filter. A study on the use of particle filters on the same context have been presented recently [32].

In [53], the authors present a multi-UAV map-building approach based on evidential reasoning. The objective is that a group of UAVs build a certainty grid about potential targets on an area. The paper shows results only in simulation, and presents an interesting comparison with a Bayesian approach to the same problem. The authors conclude that the Dempster-Shafer approach can yield better results in terms of timing when the sensors' accuracy is low. Also, the paper considers cooperative path planning methods based on the results from both, the Bayesian and evidential approaches. Nevertheless, the algorithm presented is purely centralized, and nothing is said about a decentralized version for the evidential approach.

In [7] the feasibility of the application of a team of small (low-altitude short endurance) UAVs to cooperatively monitor and track the propagation of large forest fires is explored. The paper provides simulations using a six degree of freedom dynamic model for the UAV and a numerical propagation model for the forest fire.

4.2 Probabilistic Algorithms for Cooperative Perception

As said before, the robots will employ probability theory to represent their current knowledge. The environment and the robots are characterized by what is called the *state*, which is represented at a given instant t as the vector $\mathbf{x}_t$. The objective of a *cooperative* perception system is to obtain an estimation of this state from the sensorial data on-board the different robots. $\mathbf{z}_t$ represents the set of measurements obtained at time t. In a probabilistic framework, the current knowledge about the state is given by what is called the *belief* on the state, defined as:

$$bel(\mathbf{x}_t) = p(\mathbf{x}_t|\mathbf{z}^t). \tag{4.1}$$

that is, the probability distribution on the state conditioned on all the information gathered up to time t, $\mathbf{z}^t$. From the probabilistic point of view, this conditional distribution, called the *posterior*, represents all the information the robot can compute from the sensor data collected.

The main tool for probabilistic state estimation is Bayesian inference [42, 46]. Bayesian inference allows to integrate measurements generated by the perception tools with prior knowledge about the state to obtain an updated estimation of the belief on the state. Under some assumptions, the Bayes filter allows for a recursive estimation of the state of the events. The equation for the Bayes recursion [42] is:

$$p(\mathbf{x}_t|\mathbf{z}^t) = \overbrace{\eta^{-1}p(\mathbf{z}_t|\mathbf{x}_t)}^{update} \underbrace{\int_{\mathbf{x}_{t-1}} p(\mathbf{x}_t|\mathbf{x}_{t-1})p(\mathbf{x}_{t-1}|\mathbf{z}^{t-1})\mathrm{d}\mathbf{x}_{t-1}}_{prediction} \tag{4.2}$$

$$\eta = p(\mathbf{z}_t|\mathbf{z}^{t-1}) = \int_{\mathbf{x}_t} p(\mathbf{z}_t|\mathbf{x}_t) \int_{\mathbf{x}_{t-1}} p(\mathbf{x}_t|\mathbf{x}_{t-1})p(\mathbf{x}_{t-1}|\mathbf{z}^{t-1})\mathrm{d}\mathbf{x}_{t-1}\mathrm{d}\mathbf{x}_t \tag{4.3}$$

Equation (4.2) generally is decomposed in two steps, called *prediction* and *updating*. Prediction usually implies an increase in the amount of uncertainty on $\mathbf{x}$, while updating narrows this uncertainty as a consequence of the new measurement.

The important terms in the previous expressions are the conditional distributions $p(\mathbf{x}_t|\mathbf{x}_{t-1})$ (the *transition* model) and $p(\mathbf{z}_t|\mathbf{x}_t)$ (the *measurement* model) and the conditional dependencies among the random variables.

Using the same assumptions, it is possible to derive the Bayes filter for the recursive estimation of the belief for the full state trajectory $p(\mathbf{x}^t|\mathbf{z}^t)$ [42]:

$$p(\mathbf{x}^t|\mathbf{z}^t) = \eta p(\mathbf{z}_t|\mathbf{x}_t)p(\mathbf{x}_t|\mathbf{x}_{t-1})p(\mathbf{x}^{t-1}|\mathbf{z}^{t-1}) \tag{4.4}$$

Another convenient expression for this equation, that unrolls the previous expression until considering the *prior* information $p(\mathbf{x}_0)$, is:

$$p(\mathbf{x}^t|\mathbf{z}^t) = \eta' p(\mathbf{x}_0) \prod_{\tau=1}^{\tau=t} p(\mathbf{z}_\tau|\mathbf{x}_\tau)p(\mathbf{x}_\tau|\mathbf{x}_{\tau-1}) \tag{4.5}$$

4.2.1 Multi-robot Perception

In the case of a multi-robot fleet (for instance, a fleet of UAVs), the objective is to cooperatively estimate the state of the world (that is, the relevant information, represented by $\mathbf{x}_t$) from the measurements obtained by all the robots of the fleet.

In order to determine what information should be communicated and how this information is fused with the local knowledge of each robot, first it is analyzed the resultant knowledge in the case that all information were available at any point of the fleet. That could be considered an ideal `omniscient` situation, in which a central node gets all the available information at any time.

The measurements $\mathbf{z}_t$ are the collection of all the measurements gathered by all the sensors of the fleet of robots $\{\mathbf{z}_{j,t}, j = 1, \ldots, M_t\}$. The current measurement is given by the vector $\mathbf{z}_t^m = [\mathbf{z}_{1,t}^T, \ldots, \mathbf{z}_{M_t,t}^T]^T$. Then, the belief state for the central node is given by:

$$bel_m(\mathbf{x}_t) = p(\mathbf{x}_t|\mathbf{z}^{m,t}) = \eta p(\mathbf{z}_t^m|\mathbf{x}_t) \int p(\mathbf{x}_t|\mathbf{x}_{t-1})p(\mathbf{x}_{t-1}|\mathbf{z}^{m,t-1})d\mathbf{x}_{t-1} \tag{4.6}$$

Given the assumption that the data gathered by the different robots at any time instant t are *conditionally independent* given the state $\mathbf{x}_t$, the previous equation becomes:

$$bel_m(\mathbf{x}_t) = \eta \prod_{j=1}^{M_t} p(\mathbf{z}_{j,t}|\mathbf{x}_t) \int p(\mathbf{x}_t|\mathbf{x}_{t-1})p(\mathbf{x}_{t-1}|\mathbf{z}^{m,t-1})d\mathbf{x}_{t-1} \tag{4.7}$$

For data fusion purposes, it will be very important to determine the likelihood function $p(\mathbf{z}_{j,t}|\mathbf{x}_t)$ for every source of measurements within the robots of the fleet. Likewise, in the multi-robot case, (4.5) becomes:

$$bel_m(\mathbf{x}^t) = p(\mathbf{x}^t|\mathbf{z}^{m,t}) = \eta' p(\mathbf{x}_0) \prod_{\tau=1}^{\tau=t} [\prod_{j=1}^{M(\tau)} p(\mathbf{z}_{j,\tau}|\mathbf{x}_\tau)]p(\mathbf{x}_\tau|\mathbf{x}_{\tau-1}) \tag{4.8}$$

The computation of (4.7) and (4.8) are the ideal objective for the cooperative perceptions algorithms. It should be noted that nothing has been commented yet about what the state variables are, the particular distributions considered and so forth.

4.2.2 Semi-decentralized Belief Computation

In general, it is not possible to dispose of all the data in a central node without delays due to bandwidth limitations, bounded communication ranges, etc. On the other hand, if the robots have decisional capabilities, then they should maintain their own local belief states that will be used by those decisional layers for planning activities and so forth. Therefore, the idea would be to combine in some way these local estimations, by communicating high-level belief states instead of raw data.

In a fully decentralized approach, the robots *share* their beliefs with their neighbors. Then, the received information is locally fused in order to improve the local perception of the world. The main question is if there is a way of combining the belief states so that the final belief state is closer (ideally the same) to the global belief that could be computed in the centralized case, represented by (4.7) and (4.8).

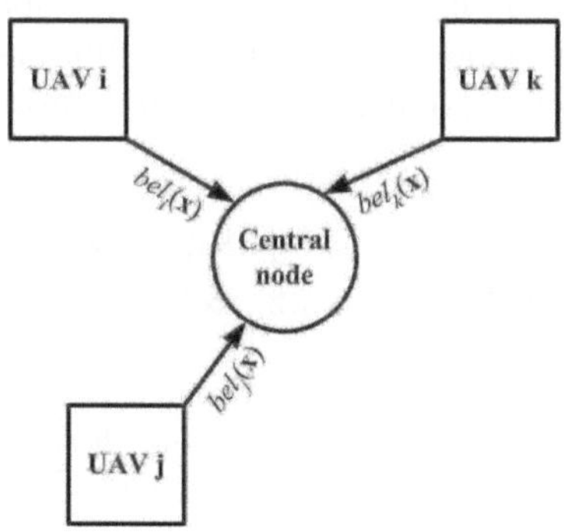

Fig. 4.2. Scheme that shows the situation in the semi-decentralized scheme

Before getting into the issues related to the fully decentralized computation, this section analyzes a first approach, depicted in Fig. 4.2. This approach will be named semi-decentralized. In it, each robot i maintains its local belief state $bel_i(\mathbf{x}_t)$:

$$bel_i(\mathbf{x}_t) = p(\mathbf{x}_t|\mathbf{z}_i^t) = \eta_i p(\mathbf{z}_{i,t}|\mathbf{x}_t) \int p(\mathbf{x}_t|\mathbf{x}_{t-1}) p(\mathbf{x}_{t-1}|\mathbf{z}_i^{t-1}) \mathrm{d}\mathbf{x}_{t-1} \tag{4.9}$$

where $\eta_i^{-1} = p(\mathbf{z}_{i,t}|\mathbf{z}_i^{t-1})$. Then, the robot sends its belief state to a central node that combines all the local beliefs to obtain a global one.

The Case of Static State

If the state is static, then the belief state at time t using all robot data is given by:

$$bel_m(\mathbf{x}_t) = \eta_m p(\mathbf{x}) \prod_{\tau=1}^{\tau=t} p(\mathbf{z}_\tau^m|\mathbf{x}) \tag{4.10}$$

where the time index has been removed to indicate that the state is static, and then $bel(\mathbf{x}_t)$ means the belief state after all the data gathered up to time t has been integrated. Similarly, for any robot i:

$$bel_i(\mathbf{x}_t) = \eta_i p_i(\mathbf{x}) \prod_{\tau=1}^{\tau=t} p(\mathbf{z}_{i,\tau}|\mathbf{x}) \tag{4.11}$$

If M robots are considered, as $p(\mathbf{z}_t^m|\mathbf{x}) = \prod_{i=1}^{M} p(\mathbf{z}_{i,t}|\mathbf{x})$, then, if the prior beliefs $p(\mathbf{x})$ are the same:

$$bel_m(\mathbf{x}_t) = \eta p(\mathbf{x}) \prod_{i=1}^{M} \frac{bel_i(\mathbf{x}_t)}{p(\mathbf{x})} \tag{4.12}$$

This equation gives a basic formula to combine the robots beliefs in order to obtain the global one. It means that the central node directly combines all the beliefs received, after removing the common information that all robot share (the prior $p(\mathbf{x})$). Another convenient way of representing the previous relations are in recursive form.

$$bel_m(\mathbf{x}_t) = \eta_m p(\mathbf{z}_t^m|\mathbf{x}) bel_m(\mathbf{x}_{t-1}) \tag{4.13}$$

$$bel_i(\mathbf{x}_t) = \eta_i p(\mathbf{z}_{i,t}|\mathbf{x}) bel_i(\mathbf{x}_{t-1}) \tag{4.14}$$

so

$$bel_m(\mathbf{x}_t) = \eta bel_m(\mathbf{x}_{t-1}) \prod_{i=1}^{M} \frac{bel_i(\mathbf{x}_t)}{bel_i(\mathbf{x}_{t-1})} \tag{4.15}$$

Expressing the beliefs in logarithmic form:

$$bel_m(\mathbf{x}_t) = \log(\eta) + bel_m(\mathbf{x}_{t-1}) + \sum_{i=1}^{M} \left[bel_i(\mathbf{x}_t) - bel_i(\mathbf{x}_{t-1})\right] \tag{4.16}$$

Therefore, what the central node has to do is to sum (in logarithmic form) into a running total the increase in *evidence* given by every robot of the fleet. Even though no particular form or representation for the belief states is assumed yet, an interesting characteristic of this case is that, as the state is static, the message size for communicating the beliefs is fixed. Moreover, the number of messages sent by each robot does not depend on the number of robots [33]. Finally, one of the most important issues is that, in this case, the beliefs can be received *asynchronously* and with *arbitrary latency* (i.e., in the previous equations $t-1$

can be substituted by the previous instant in which the central node received information from each robot). Hence, in the case of a static state, each robot can accumulate evidence and transmit it at convenience, without additional memory costs.

Dynamic Environments

In order to reconstruct the ideal centralized belief state, as noted in [33], in the dynamic case the role that was played before by belief state at time t, $bel(\mathbf{x}_t)$, is played now by the belief state over the full state trajectory up to time t, $bel(\mathbf{x}^t)$.

The belief state over the full trajectory for robot i is:

$$bel_i(\mathbf{x}^t) = p(\mathbf{x}^t|\mathbf{z}_i^t) = \eta' p(\mathbf{x}_0) \prod_{\tau=1}^{\tau=t} p(\mathbf{z}_{j,\tau}|\mathbf{x}_\tau) p(\mathbf{x}_\tau|\mathbf{x}_{\tau-1}) \tag{4.17}$$

Comparing this expression to (4.8) then:

$$bel_m(\mathbf{x}^t) = \eta p(\mathbf{x}_0^t) \prod_{i=1}^{M} \frac{bel_i(\mathbf{x}^t)}{p(\mathbf{x}_0^t)} \tag{4.18}$$

where $p(\mathbf{x}_0^t) = p(\mathbf{x}_0) \prod_{\tau=1}^{\tau=t} p(\mathbf{x}_\tau|\mathbf{x}_{\tau-1})$. Therefore, if each robot sends its belief state over the state trajectory, the beliefs can be combined to obtain a global one that would be equal to the one obtained in a centralized version. Moreover, as in the static case, the belief states can be received asynchronously. Each robot can accumulate evidence, and send it whenever is possible to communicate with the central node. However, the problem in this case is that the state grows over time, and therefore the size of the message needed to communicate the corresponding beliefs. In general, this will make this scheme of communication unaffordable. Nevertheless, for the normal operation of the robots, only the state trajectory over a time interval is needed, so this belief trajectory can be bounded. In any case, depending on the representation of the belief states and the dimension of the state itself, the size of information needed to be transmitted can be prohibitively high.

4.2.3 Decentralized Cooperative Perception

In a fully decentralized scheme, there is no central node that fuses the beliefs obtained by the robots. In a decentralized fleet, no robot have knowledge of the global network topology, and there is no global communication facility, in the sense that cannot be ensured that the information transmitted by one robot will reach all the robots of the fleet [16]. At any time instant, each robot will be able to communicate and share its belief directly with a subset of the fleet.

There is a situation in which the previous general equations can be applied. If the belief network topology[1] is a tree as shown in Fig. 4.3, that is, if there

[1] The term network refers in this case should be interpreted at the data level. That is, the underlying physical or transportation layers may have a different topology.

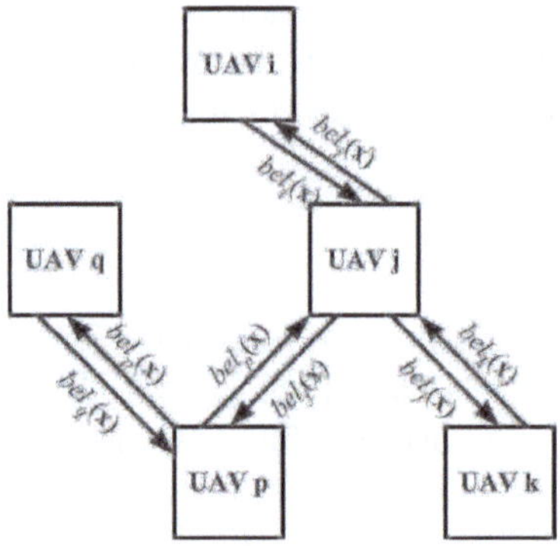

Fig. 4.3. A group of UAVs forming a tree-shaped network

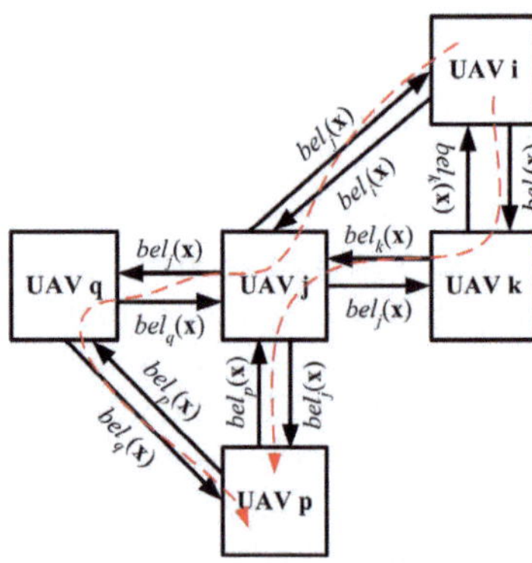

Fig. 4.4. Multiple paths can lead to overestimation, when incorporating several times the same information

is a unique path between any pair of providers and receivers, then the fusion equations (4.12) and (4.18) for the central node can be applied at every robot to fuse the beliefs received from its neighbors [49], and then update their own beliefs. The only aspect that must be considered is to avoid considering multiple times the same information through the direct link between two robots with direct communication (like UAV p and UAV q in Fig. 4.3). However, this can be solved locally, without further information about the rest of the network. Each robot only needs to know what information it transmitted to its neighbors in order to remove it when integrating received data.

However, the same scheme cannot be applied to general topologies without further considerations. Consider the scheme of connections of Fig. 4.4. In this case, there are multiple paths from the same UAV to others. The problem in this case is that the *same information* could be accounted *several times*. Thus, the belief combination rule has to consider the common information [16, 22, 45]. Otherwise, this could lead to overconfident estimates.

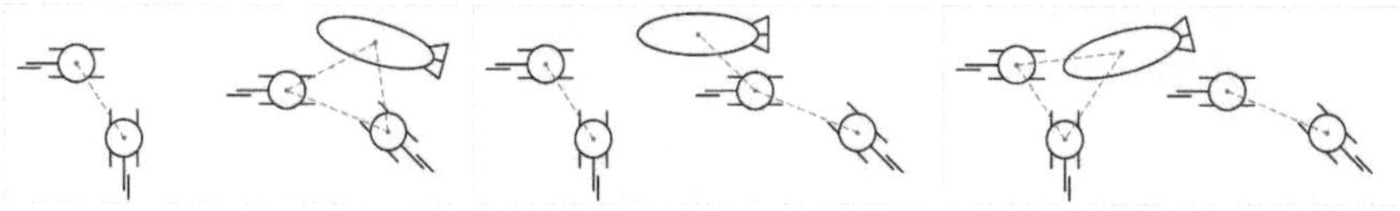

Fig. 4.5. In the general case, the topology of the network will change with time

Without additional assumptions on the form of the belief distributions, their representations and others, there is no general solution to the problem of considering the common information in networks of arbitrary topology [49], if only local information is considered. One solution would be to force a tree topology in the network. However, in a fleet of robots, the topology of the connections is dynamic and the tree structure of the network would have to be ensured along the time. Also, if a tree topology is forced, an interesting property is lost. One robot can act as a *data mule* between different subsets of robots of the fleet (see Fig. 4.5). When connected to one subset, the robot updates its belief from the information received from the subset. Then, when connected with the second subset, the robots of this second subset will incorporate information from the other subset indirectly through the belief state of the moving robot.

The previous sections have presented the general problem of information fusion for cooperative perception. The development of working algorithms, even in the case of centralized fusion, requires defining a particular representation of the belief states, which itself depends on the problem.

4.2.4 Developing Cooperative Perception Actions

The previous sections have showed the main issues in decentralized information fusion. The final important aspect of cooperative perception is the ability to develop actions to improve the knowledge about the environment. The objective is to determine which actions $\mathbf{u}^m$ the robots should carry out in order to increase some measurement about the goodness of the current information.

Determining robots actions considering uncertain worlds is usually tackled employing decision theoretic techniques. In the previous equations of the Bayes filter, the actions of the robots $\mathbf{u}^m$ can be considered explicitly.

$$bel_m(\mathbf{x}_t) = \eta \prod_{j=1}^{M_t} p(\mathbf{z}_{j,t}|\mathbf{x}_t) \int p(\mathbf{x}_t|\mathbf{x}_{t-1}, \mathbf{u}_t^m) p(\mathbf{x}_{t-1}|\mathbf{z}^{m,t-1}, \mathbf{u}^{m,t-1}) \mathrm{d}\mathbf{x}_{t-1} \tag{4.19}$$

The objective is to select actions considering the current knowledge of the world (represented by the belief state) and the possible outcomes from actions carried by the robot. The computation of rational actions is considered including a payoff or reward function on the state $R(\mathbf{x}), \mathbb{R}^n \longrightarrow \mathbb{R}$. This payoff value can consider costs and revenues for being in a particular state. The objective is to find an optimal *policy*, that is, a function $\pi(\mathbf{x}) \rightarrow \mathbf{u}$, that determines what action $\mathbf{u}$ should be carried out if the state is $\mathbf{x}$. The policy should be optimal in the sense that it should maximize the expected cumulative payoff (also called expected utility):

$$\pi^* = \arg\max_{\pi} E\left[\sum_{t=0}^{T} \gamma^t R(\mathbf{x}_t)|\pi\right] \tag{4.20}$$

that this, the expected sum of the rewards from time 0 to time T (T is called the planning horizon and $\gamma \in [0,1]$ is the discount factor).

If the environment is fully observable (that is, if we are certain about the state any time we get new information), but the effect of the actions carried out is non-deterministic, the problem is called a *Markov Decision Process* (MDP) [34]. If the more general case in which the environment is not observable, the problem is called *Partially Observable Markov Decision Processes* (POMDPs)[40].

Choosing a convenient payoff function, POMDPs can be used for developing perception actions; in this case, the payoff function usually considers higher values for narrower belief states, meaning that more informative belief states are preferred. However, the POMDP framework is usually infeasible in this case, as the number of unknown is high. The problem is even more important in the case of multiple robots, in which the problem to be solved is to determine an optimal policy that computes the actions of the robot of the fleet $\mathbf{u}^m$ given the global belief state $bel_m(\mathbf{x})$.

Another option for developing cooperative perception actions is to define a measurement of the information gain obtained when executing a certain information gathering task. Several different measurements of the information gain can be used. For instance, for unimodal distributions, the covariance could give a measure of the amount of information gained when executing an action (the bigger the covariance, the more uncertain about the actual value of the state). Another and more general measure about the information is the entropy of a probability distribution. The entropy $H(p)$ of a probability distribution $p(x)$ is defined as the expected value of the information $-\log[p(x)]$. That is, the entropy of the belief state is given by:

$$H(t) = E[-\log bel(\mathbf{x}_t)] = -\int bel(\mathbf{x}_t)\log bel(\mathbf{x}_t)d\mathbf{x}_t \tag{4.21}$$

The information gain is defined as the variation in the entropy after carrying an action $\mathbf{u}_{t+1}$. When executing this action, a new belief state $bel(\mathbf{x}_{t+1})$ will be obtained from the measurement $\mathbf{z}_{t+1}$ received, with a new entropy $H(\mathbf{u}_{t+1}, \mathbf{z}_{t+1})$. However, only the action $\mathbf{u}_{t+1}$ can be selected. Therefore, what can be computed is the expected entropy considering the potential measurements $H(\mathbf{u}_{t+1}) = E_{\mathbf{z}_{t+1}}[H(\mathbf{u}_{t+1}, \mathbf{z}_{t+1})]$. Therefore, the information gain associated to action $\mathbf{u}_{t+1}$ is defined as follows:

$$I(\mathbf{u}_{t+1}) = H(t) - E_{\mathbf{z}_{t+1}}[H(\mathbf{u}_{t+1}, \mathbf{z}_{t+1})] \tag{4.22}$$

This metric can be used to establish preferences among actions. A policy that at the same time maximizes the information gain and minimices action costs can be used for developing cooperative actions for perception purposes.

The decisional architecture for multi-robot coordination presented in Chap. 2 can consider coordinated information gathering tasks, as cooperative detection and confirmation, and cooperative monitoring. Nevertheless, the previous considerations are not yet included and the actions are selected following off-line learned models of performance (like detection rates and precision in localization tasks).

4.3 Vision-Based Object Detection, Localization and Tracking with Multiple UAVs

4.3.1 Initial Assumptions

In order to test the decentralized perception scheme presented above, the application to the detection, localization and tracking of events of interest by a fleet of UAVs is considered. The objective of the fleet is to detect potential targets, estimate their positions and also discard false alarms.

The important information for this scenario is the location of the UAVs and the location (and perhaps other information) of the events of interest. Then, the state to be estimated in this case, in its most simple version, is comprised by the information associated to a set of N events and by the position of the M UAVs.

It is assumed that the UAVs carry on-board Global Positioning System (GPS) receivers, Inertial Measurement Units (IMUs) and other sensors, different from cameras, for navigation. These sensors are used to localize the UAVs in a common global frame, and thus the localization problem is not considered here. However, as it will be seen, the conditional dependence of the measurements obtained by the cameras on the robot pose estimates (denoted by $\mathbf{q}_t$) must be taken into account.

The second assumption is that the actions of the robots will not affect the state of the objects to be tracked. This assumption usually holds when the objects to be tracked are not aware of the robots patrolling, for instance, if the objects of interest are fire alarms, or are not *evading* targets, but this would not hold for evading targets or opponent robots [51].

4.3.2 Event State Definition

In the general case of event detection and localization, the state $\mathbf{x}$ to be tracked obviously include the position of the object $\mathbf{p}_t$. If a moving object is considered, the velocity $\dot{\mathbf{p}}_t$ is also included into the state to be estimated. This will be called the kinematic part of the state.

Further information will be also needed. An important objective in some missions is to confirm that an object belongs to a certain class within a set $\mathbf{\Gamma}$ (for instance, in the case of fire alarms detection, this set will include as classes fire alarms and false alarms). Therefore, the state will include information regarding the classification of the object. Also, in certain applications, some appearance information could be needed to characterize an event, which also can help in the

Fig. 4.6. An infrared and a visual image of a fire alarm. The objective is to fuse all the data from the different vehicles of the fleet to obtain a global belief on the parameters of a given alarm.

task of data association between different UAVs with different cameras. This kind of information usually will be static, and will be represented by $\boldsymbol{\theta}$.

The complete state to be estimated is composed by the status of all the events. The number of events (N_t) can vary with the time. The state at time t is then represented by vector $\mathbf{x}_t = [\mathbf{x}_{1,t}^T, \ldots, \mathbf{x}_{N_t,t}^T]^T$. Each potential alarm i is defined by:

$$\mathbf{x}_{i,t} = \begin{pmatrix} \mathbf{p}_{i,t} \\ \dot{\mathbf{p}}_{i,t} \\ \boldsymbol{\theta}_i \end{pmatrix}. \tag{4.23}$$

4.3.3 The Likelihood Function for Vision

The key point in the Bayesian framework adopted is to determine the likelihood function $p(\mathbf{z}_t|\mathbf{x}_t, \mathbf{q}_t)$ for the case of the cameras. Notice that the (uncertain) pose of the UAV is included in the model, and to determine the final likelihood this dependence will be marginalized out.

In the limit, $p(\mathbf{z}_t|\mathbf{x}_t, \mathbf{q}_t)$ would be a probabilistic model of image formation, that is, the probability that the pixels of the image have certain illuminance or colour values given the status of the environment and the position of the sensor. In most cases it is not needed such a complex model. Instead of the raw pixels, some features related to the application are extracted from the images. Nevertheless, it is required to formulate probabilistically all these steps in order to incorporate the measurements in the estimation process.

Feature Extraction: Segmentation Algorithms

In the application considered (object detection and localization), the images captured by the cameras on-board the UAVs should be analyzed looking for the objects of interest. The image processing functions should be able to segment the objects of interest on the image plane, and differentiate them from the background. Moreover, the algorithms will obtain a set of features related to the identity of the object $\boldsymbol{\theta}$. In order to include this into the probabilistic framework, it is needed to relate the visual features to the state through:

$$p(\mathbf{z}_t|\mathbf{x}_t) = p(\mathbf{z}_t|\mathbf{p}, \dot{\mathbf{p}}, \boldsymbol{\theta}), \boldsymbol{\theta} \in \boldsymbol{\Gamma} \tag{4.24}$$

These features will depend on the application considered. In general, determining the likelihood function will consist on a learning phase over the sensorial space; this learning phase should provide a map between features and classes.

In the particular application of fire detection, the space state for the identity part of the state $\boldsymbol{\theta}$ consists only on two potential values, $\boldsymbol{\Gamma} = \{\text{fire, no fire}\}$. Chapter 8 will present the techniques employed for forest fire detection using cameras (of different modalities, both infrared and visual). These algorithms provide directly a binary decision over the image plane, so that the measurements $\mathbf{z}_t$ are a set of blobs over the image plane classified of fire. As it will be seen, it is important also to employ the negative information encoded in the regions of the image not classified as fire.

There is always the chance for *false positives* and *misdetections*. False positives occur when the algorithm detects objects but there are no objects of the considered class in the field of the camera. Misdetections happen if no response is given when an object is present. These two facts have to be used to determine the likelihood function associated to the processed images.

A simple model will be used. It consists of characterizing the segmentation algorithms by two values:

- The probability of detection (P_{D}), defined as the likelihood that, being an object of the given class within the field of view, the object is effectively detected $p(\mathbf{z}_t|\mathbf{p}, \dot{\mathbf{p}}, \boldsymbol{\theta} = \text{fire})$.
- The probability of false positive (P_{F}), defined as the likelihood that the algorithm generates a response when no actual object is within the field of view of the camera $p(\mathbf{z}_t|\mathbf{p}, \dot{\mathbf{p}}, \boldsymbol{\theta} = \text{no fire})$.

Projective Geometry

In order to complete the main aspects of the likelihood function, it is needed to relate objects on the image plane with the position of objects in the 3D world (the location part of the state $\mathbf{p}_t$). Cameras project points in the space into points on the image plane. Cameras are usually modeled using the tools of projective geometry [18]. The projection is modeled by the *pin-hole* projection model. Following this model, each point in the space, $\mathbf{p}$ and its corresponding image pixel $\mathbf{m}$ on the image plane of the camera are related by (4.25), where $\mathbf{p}$ and $\mathbf{m}$ are in homogeneous coordinates:

$$s\mathbf{m}_t = \mathbf{A}\left(\mathbf{R}_t \; -\mathbf{t}_t\right)\mathbf{p}_t \tag{4.25}$$

where $\mathbf{A}$ is the upper triangular internal calibration matrix of camera. $\mathbf{R}_t$ and $\mathbf{t}_t$ are the rotation and translation that refer the camera coordinate system and a global reference frame, and are part of the estimated state of the UAV $\mathbf{q}_t$. Several

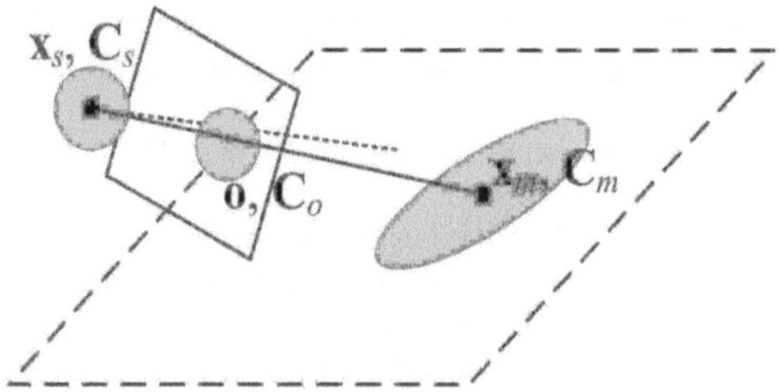

Fig. 4.7. Propagation of uncertainties through the geolocation procedure

methods can be used for camera calibration. Here, the method presented by [55] is used.

Equation (4.25) implies a non-linear relation between the state and the measurements. Moreover, if the pose is uncertain, it has to be considered when obtaining the corresponding likelihood.

Putting together the geometric model of the camera with the probabilistic characterization of the segmentation algorithms allow to establish a likelihood function for the image-base measurements obtained by the UAVs. $p(\mathbf{z}_t|\mathbf{x}_t, \mathbf{q}_t)$ would be the probability of getting the features extracted (that is, the objects segmented on the image plane) given the state of the events $\mathbf{x}_t$ and the position and orientation of the camera $\mathbf{q}_t$.

Prior Position Information from Cameras

As it will be seen, sometimes it is needed to determine the position of an object of interest from its position on the image plane. For instance, when an object is detected by the first time, one should obtain an initial estimation of its position $\mathbf{p}$.

Unfortunately, relation (4.25) is not invertible. Therefore, although $\mathbf{A}$, $\mathbf{R}$ and $\mathbf{t}$ may be known, from the position on the image plane of an object $\mathbf{m}$, it is not possible to recover its 3D position $\mathbf{p}$. If nothing more is known, cameras provide only bearing information, and the full 3D position is not observable.

Several images from different points of view of the same object can be used to estimate the range of a point, and thus its 3D position, by triangulation, as in stereo vision [3, 9]. Also, if a Digital Elevation Map (DEM) of the scene is available, and the pose of the camera is known in the same coordinate frame than the DEM, then it is possible to estimate the position of an object on the ground by ray tracing techniques, provided that the camera is calibrated ($\mathbf{A}$ is known). Through this Chapter, it is assumed that such a map exists. In some applications, this kind of map could be provided by an UAV commanded to generate a DEM of the scenario [20]. Using the geolocation procedure, the UAVs will provide as measurements direct estimations on the position of potential alarms.

4.4 Local UAV Image Processing Techniques

The different UAVs will process locally their cameras in order to detect and track events. Interesting events could be mobile objects on a zone, fire alarms, and others. This section focuses on feature extraction and tracking and other set of basic functionalities local to each UAV, while Chapter 8 will show the fire segmentation techniques. These local functionalities are needed for cooperative detection and monitoring activities. The algorithms refer to vision, which currently is the most used exteroceptive sensor on-board UAVs.

4.4.1 Image Motion Estimation

Many algorithms and applications require as input an estimation on the image motion (for instance, mobile object detection or motion compensation, as it will be seen). Dense motion fields (or optical flow) has been used for monitoring activities using UAVs [8]. However, the computation of dense motion fields usually requires small displacement between consecutive images, which usually is not the case when an UAV is moving. Here, a method based on matching point-like features allows to obtain a sparse image motion field under less constrained motion. The image matching method presented is related to the described in [12], although significant improvements have since been made, as summarized in [30].

In [12] corner points were selected using the criteria described in [48]; each point was the centre of a fixed-size window which is used as a template in order to build matching window sequences over the stream of video images. Window selection provides for initial startup of window sequences as well as candidates (called direct candidates) for correlation-based matching tries with the last known template window of a sequence. The selection of local maxima of the corner detector function assured stable features, so window candidates in any given image were usually near the right matching position for some window sequence.

The correlation-based matching process with direct candidates within a search zone allowed to generate a matching pair data base, which described possibly multiple and incompatible associations between tracked sequences and candidates. A disambiguation process selected the right window to window matching pairs by using two different constraints: least residual correlation error and similarity between clusters of features.

The new approach uses the same feature selection procedure, but its matching strategy is significantly different. First, the approach ceases to focus in individual features. Now clusters are not only built for validation purposes; they are persistent structures which are expected to remain stable for a number of frames, and are searched for as a whole. Second, the disambiguation algorithm changes from a relaxation procedure to a more efficient predictive approach, similar to the one used in [3] for contour matching. Rather than generating an exhaustive data base of potential matching pairs, only selected hypothesis are considered. Each hypothesis, with the help of the persistent cluster data base, allows to define reduced search zones for sequences known to belong to the same cluster as the hypothesis, if a model for motion and deformation of clusters is known.

For this approach to be feasible, the following points should be defined

- Cluster model.
- Cluster building and validation.
- Cluster based prediction.

Cluster Model

The similarity of shape between regions of different images is verified by searching for clusters of windows whose members keep the same relative position, after a scale factor is applied. For a cluster of window sequences $\Gamma = \{\Phi_1, \Phi_2, \ldots, \Phi_n\}$, the shape similarity constraint is given by the following expressions:

$$\left| \frac{\|w_{\Phi_k}, w_{\Phi_l}\|}{\|v_{\Phi_k}, v_{\Phi_l}\|} - \frac{\|w_{\Phi_p}, w_{\Phi_q}\|}{\|v_{\Phi_p}, v_{\Phi_q}\|} \right| \leq k_p \ , \forall \Phi_k, \Phi_l, \Phi_p, \Phi_q \in \mathit{\Gamma} \quad (4.26)$$

In (4.26), k_p is a tolerance factor, w_i are candidate windows in the next image and v_i are template windows from the preceding image. The constraint is equivalent to verify that the euclidean distances between windows in both images are related by a similar scale factor; thus, the ideal cluster would be obtained when euclidean transformation and scaling can account for the changes in window distribution. Furthermore, an additional constraint over the maximum difference of rotation angle between pair of windows is used:

$$|\alpha_{\Phi_k,\Phi_l} - \alpha_{\Phi_p,\Phi_q}| \leq \gamma_p \ , \forall \Phi_k, \Phi_l, \Phi_p, \Phi_q \in \mathit{\Gamma} \quad (4.27)$$

where α_{rs} is the rotation angle of the vector that links windows from sequences r and s, if the matching hypothesis is accepted, and γ_p is a tolerance factor. Although the cluster model is adequately simple and seems to fit the current applications, more realistic local models such as affine or full homography could be easily integrated in the same scheme.

Cluster-Based Prediction for Hypotesis Propagation

It is easy to verify that two hypothesized matching pairs allow to predict the position of the other members of the cluster, if their motion can be modeled approximately by euclidean motion plus scaling, as the constraints (4.26) and (4.27) imply. Using this model, the generation of *candidate clusters* for a previously known cluster can start from a primary hypothesis, namely the matching window for one of its window sequences. This assumption allows to restrict the search zone for other sequences of the cluster, which are used to generate at least one secondary hypothesis. Given both hypothesis, the full structure of the cluster can be predicted with the small uncertainty imposed by the tolerance parameters k_p and γ_p, and one or several candidate clusters can be added to a data base.

For a specific cluster, assuming that a primary hypothesis is available, the cluster candidate generation process involves three steps:

- Secondary hypothesis generation.
- Cluster-based prediction.
- Indirect candidate generation.

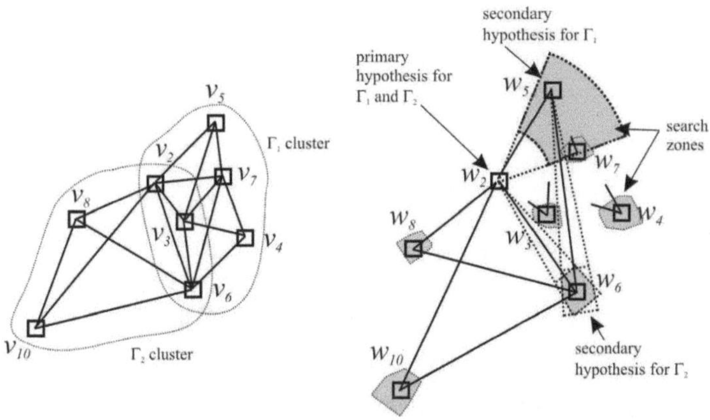

Fig. 4.8. Generation of cluster candidates

Secondary hypothesis are generated by single-window prediction. A search area can be defined by considering maximum limits of frame-to-frame cluster scaling and rotation, based on the knowledge of the environment; thus, the search of matching pairs can be restricted to the direct candidates contained in the area. Any window sequence linked to the cluster can be used as secondary hypothesis this way.

Several primary/secondary pairs can be obtained. For each available pair of primary/secondary hypothesis, a full cluster can be predicted; in this case, the tolerance parameters k_p and γ_p are used to further constraint the search zone of each component. Matching windows for each sequence can be found by exploring the set of direct or previously generated indirect candidates located within the computed limits.

If there remains some window sequences without candidate after the direct candidate expansion, new indirect candidates are generated by prediction of a suitable starting point for the error minimization algorithm.

The creation of any given candidate cluster can trigger the creation of others for neighbour clusters, provided that there is some overlap among them; in Fig. 4.8, for example, the creation of a candidate for cluster 1 can be used immediately to propagate hypothesis and find a candidate for cluster 2. Direct searching of matching windows is thus kept to a minimum. Indeed, it makes sense to consider a higher likelihood for candidates which already take part in a sufficiently large cluster candidate.

As Fig. 4.9 shows, successful candidate generation can be propagated through neighbor clusters. Two clusters can be said to be connected if they can be linked by an arbitrary long chain of overlapping clusters; the set of clusters can be partitioned in subsets of connected elements, or groups. The computational efficiency of the process is increased by maximizing propagation, so that candidates are reused whenever possible and direct search and error-minimization is avoided. In addition, window candidates which are already supported by a cluster candidate

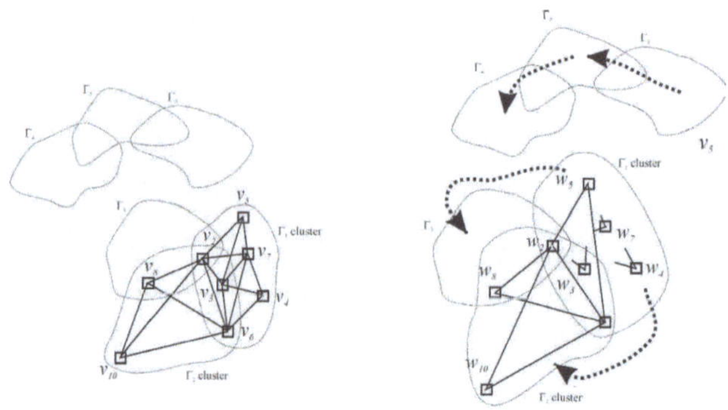

Fig. 4.9. Hypothesis propagation through cluster groups

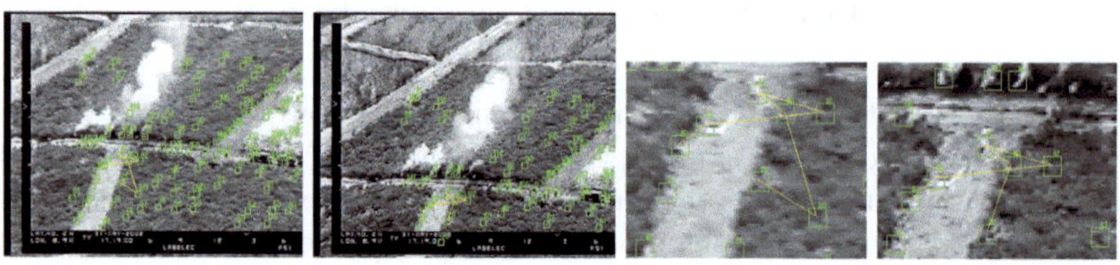

Fig. 4.10. Example of cluster matching

should be preferred to unsupported candidates. At the final stage of the method, the best cluster candidates are used to generate clusters in the last image, and determine the matching windows for each sequence. The practical result of the approach is to drastically reduce the number of matching tries, which are by far the main component of processing time when a great number of features have to be tracked, and large search zones are needed to account for high speed image plane motion. This is the case in non-stabilized aerial images, specially if only relatively low frame rate video streams are available. Figure 4.10 shows an example with two frames of an aerial sequence.

4.4.2 Blob Features

Although point-like features are suitable for image motion estimation, for certain applications, such as matching disparate views (for instance, views taken by *different* UAVs), features with more invariance properties are needed.

Homogeneity features are called *blobs* in scale-space theory [23]. In contrast to segmentation, blob detection does not attempt to segment out exact shapes of objects, but to extract robust and repeatable features discarding exact shapes and thin connections between patches (see Fig. 4.11). Blob features are also related to *maximally stable extremal regions* (MSER) [29]. MSER features are regions grown around an intensity extrema (maxima or minima) and are used to generate affine invariant frames, which are then used for view based object recognition [29].

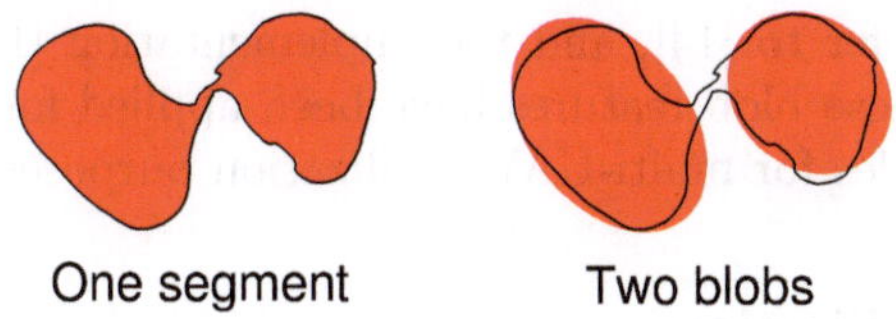

Fig. 4.11. Difference between segmentation and blob detection

Blob features can be extracted using a clustering pyramid built using robust estimation in local image regions [14, 13]. Each extracted blob is represented by its average colour $\mathbf{p}_k$, area a_k, centroid $\mathbf{m}_k$, and inertia matrix $\mathbf{I}_k$, i.e. each blob is a 4-tuple

$$\mathcal{B}_k = \langle \mathbf{p}_k, a_k, \mathbf{m}_k, \mathbf{I}_k \rangle \ . \tag{4.28}$$

Since an inertia matrix is symmetric, it has 3 degrees of freedom, and we have a total of $3+1+2+3=9$ degrees of freedom for each blob. Figure 4.12 gives a brief overview of the blob detection algorithm.

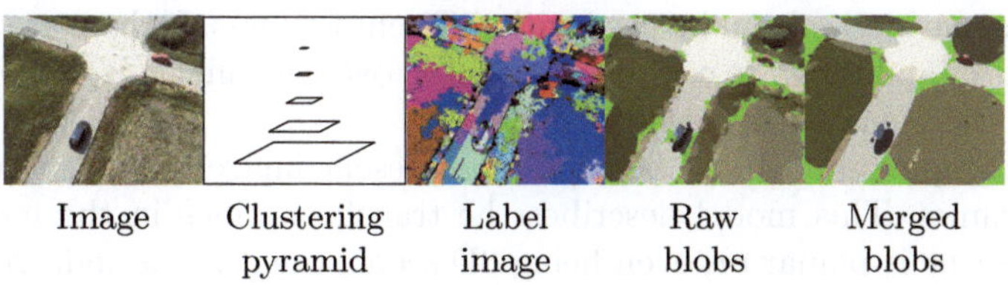

Fig. 4.12. Steps in blob detection algorithm

Starting from an image, the algorithm constructs a clustering pyramid, where a pixel $\mathbf{p}^*$ at a coarser scale is computed as the robust average of 12 pixels $\mathbf{p}_k$ at the lower scale

$$\arg\min_{\mathbf{p}^*} \sum_k r_k \rho(||\mathbf{p}_k - \mathbf{p}^*||) \, . \tag{4.29}$$

Regions where the support of the robust average is below a threshold $c_{\min}$ have their confidence flag r^* set to zero. The algorithm then creates a label image using the pyramid, by traversing it top-down, assigning new labels to points which have their confidence flag set, but do not contribute to any robust mean on the level above. The labelling produces a set of compact regions, which are then merged in the final step by agglomerative clustering. Regions with similar colour, and which fulfil the condition

$$M_{ij} > m_{\text{thr}} \sqrt{\min(a_i, a_j)} \tag{4.30}$$

are merged. Here M_{ij} is the count of pixels along the common border of blobs $\mathcal{B}_i$ and $\mathcal{B}_j$, and m_{thr} is a threshold. For more information about the feature estimation, please refer to [14], and the implementation [1], both of these are available on-line. These blob features have been applied for view matching between different vehicles for multi-UAV localization purposes [26].

4.4.3 Image Stabilization

Monitoring activities are easier if the visualized scene is static; also, the processing algorithms are simpler because the reference positions are fixed along the digital process.

Considering UAVs with hovering capabilities, unavoidable control errors, turbulence and vibrations produce changes in the camera position which lead to image motion. Therefore, for certain applications it is necessary to solve the camera motion problem. Currently, electro-mechanic systems can solve this problem, but they are heavy, expensive and usually have a residual vibration.

The proposed technique to cancel the camera motion is based on the apparent motion computation between two consecutive image frames. The results presented in this section are obtained by using the robust point-like feature tracking method presented in Sect. 4.4.1, which provides a sparse image motion field. The same method can be used with different features. If there are enough of these features and they are equally distributed over the image, it will be possible to extract the scene apparent motion.

A homographic model will be used to represent image motion between consecutive frames. This model describes the transformations in the image plane when the scene is planar or, even being 3D scene, the camera undergoes a pure rotation (which can model the motion of a camera affected by vibrations).

The algorithm described here assumes that the percentage of objects with independent motion in the scene is low, and can be treated as outliers (objects with independent motion can mask the fundamental scene movement generated by vibrations or camera motion). The algorithm should be able to detect those outliers.

Finally, a homographic model is fitted to the computed sparse motion field between two consecutive images. Then, the inverse model is applied to all pixel positions in the current image to compensate the motion, this process is called *image warping*.

Homography Computation

A homography is any linear and invertible application from the projective space $\mathbf{P^2}$ into $\mathbf{P^2}$ [18]. These applications transform lines into lines. It is represented by a 3×3 invertible matrix defined up to scale factor, so it is composed by eight independent parameters.

If a set of points in the scene lies in a plane, and they are imaged from two viewpoints, then the corresponding points in images i and j are related by a plane-to-plane projectivity or planar homography, $\mathbf{H}$:

$$s\tilde{\mathbf{m}}_i = \mathbf{H}\tilde{\mathbf{m}}_j \tag{4.31}$$

where $\tilde{\mathbf{m}}_k = [u_k, v_k, 1]^t$ is the vector of homogenous image coordinates for a point in image k, $\mathbf{H}$ is a homography matrix and s is a scale factor. The same equation holds if the image to image camera motion is a pure rotation. Even though the hypothesis of planar surface or pure rotation may seem too restrictive, they have proved to be frequently valid for aerial images ([30, 6] show some examples). An approximate planar surface model usually holds if the UAV flies at a sufficiently high altitude, while an approximate pure rotation model is a good approximation for the motion induced by a hovering helicopter. Thus, the computation of $\mathbf{H}$ will allow under such circumstances to compensate for camera motion.

Since $\mathbf{H}$ has only eight degrees of freedom, we only need four matches to determine $\mathbf{H}$ linearly. In practice, more than four correspondences are available, and the overdetermination is used to improve accuracy. For a robust recovery of $\mathbf{H}$, it is necessary to reject outlier data.

The homography computation is summarized in Fig. 4.13. It can be seen that the computation is divided in two basic steps: *oulier rejection* and *robust estimation*. The first step tries to detect the outliers in order to increase the accuracy of the computed homography. In the proposed applications, outliers will not always be wrong matching pairs; image zones where the homography model does not hold (moving objects, buildings or structures which break the planar hypothesis) will also be regarded as outliers, although they may offer potentially useful information. The overall design of the outlier rejection procedure used in this work is based on LMedS [54].

In the second step, *robust estimation*, the homography is computed by using an iterative method (M-Estimator) that allows to automatically weight the set of data in function of the residue in each iteration. It was selected a simple weighting function, concretely a Fair function, because the set of data were well filtered by the outlier rejection stage. The Fair M-Estimator will guarantee a good convergence to the correct solution in few iterations.

Image Warping

Given the homography that relates two images, a model with information about the motion of each pixel is known. This model can be used to compensate the existing motion in order to avoid camera vibrations or movements.

Thus, if $\mathbf{H}$ is the homography that represents the motion produced from image I to image I', the new position of the pixels of I after the compensation will be:

$$\mathbf{H} = \begin{bmatrix} h_{11} & h_{12} & h_{13} \\ h_{21} & h_{22} & h_{23} \\ h_{31} & h_{32} & h_{33} \end{bmatrix}, \quad \begin{array}{rl} k = & x * h_{31} + y * h_{32} + h_{33} \\ x' = & (x * h_{11} + y * h_{12} + h_{13})/k \\ y' = & (x * h_{21} + y * h_{22} + h_{23})/k \end{array} \tag{4.32}$$

In general, the transformed position $\mathbf{m}' = [x', y']^t$, corresponding to pixel $\mathbf{m} = [x, y]^t$, will not be an integer due to the algebraic operations.

It is necessary to define a method to assign an integer position to the transformed pixels in order to warp correctly the images. This problem can be seen in Fig. 4.14, where the point marks the position where the pixel has been placed.

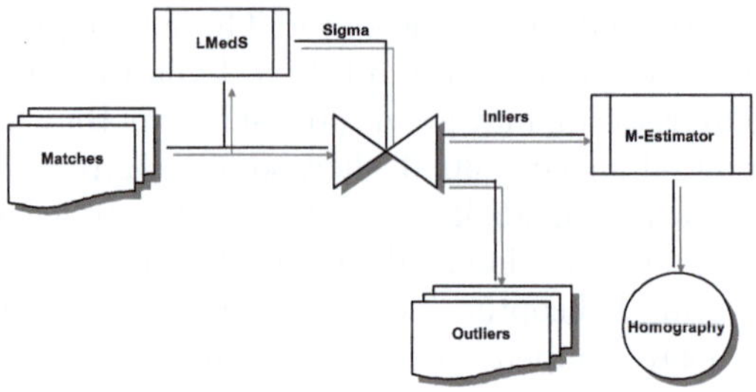

Fig. 4.13. Homography computation diagram

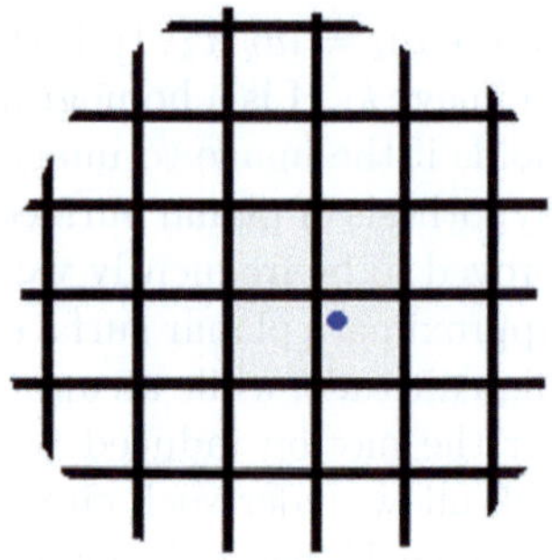

Fig. 4.14. Warping problem representation

Fig. 4.15. The sequences show a moving car and other UAV of the fleet detected using a motion detection algorithm based on the image stabilization procedure

The integer nature of the pixels force to select one of the four point neighbors (shadow pixels in the figure).

A method based on the *pixel similarity* is used. The technique uses the euclidean distance between the RGB value of the actual pixel and the values of the four neighbors in the previous image (shadow pixels in Fig. 4.14). The position of the pixel will be the one corresponding to the neighbor with the shortest distance. Thus, for each pixel, the method tries to minimize the RGB differences respect to the previous image. This helps to increase the alignment between sequenced images even correcting little errors in the homography computation.

The image stabilization procedure described above can be used for moving objects detection using monocular sequences of images. Given two consecutive images, and using the stabilization procedure, the two images can be warped to the same image coordinate frame. There, moving regions can be detected analyzing image differences. The moving regions will be due to independently moving objects or fixed objects with parallax respect to the plane used as reference for the warping procedure. Figure 4.15 shows some results.

4.5 Information Filter-Based Multi-UAV Detection and Localization Using Vision

The first approach presented in this Chapter employs Gaussian distributions to represent the belief state. They can be employed to represent unimodal beliefs, with a value with maximum probability (the mean) and uncertainty represented by the covariance of the distribution. They are used elsewhere to represent beliefs about positions of objects, etc.

Therefore, the prior knowledge at some time will be described by a Gaussian distribution $p(\mathbf{x}_0) \sim \mathcal{N}(\boldsymbol{\mu}_0, \boldsymbol{\Sigma}_0)$. If the state at time t is a linear function of the state at time $t-1$ plus Gaussian noise $\boldsymbol{\nu}_t \sim N(\mathbf{0}, \mathbf{R}_t)$, and the measurements are a linear function of the state plus Gaussian noise $\boldsymbol{\varepsilon}_t \sim N(\mathbf{0}, \mathbf{S}_t)$:

$$\mathbf{x}_t = \mathbf{A}_t \mathbf{x}_{t-1} + \boldsymbol{\nu}_t, \tag{4.33}$$

$$\mathbf{z}_t = \mathbf{M}_t \mathbf{x}_t + \boldsymbol{\varepsilon}_t, \tag{4.34}$$

then the Bayes filter (4.2) reduces to the well-known Kalman filter.

In a multi-robot application, the Information form of the Kalman filter or Information Filter (IF) [45, 46] has some interesting properties. Information filters are derived employing the canonical form of Gaussian distributions. In the canonical form, the parameters of a Gaussian distribution are the information vector $\boldsymbol{\xi}$ and the information matrix $\boldsymbol{\Omega}$. If $\boldsymbol{\mu}$ and $\boldsymbol{\Sigma}$ are the mean and covariance matrix of a Gaussian distribution, its information matrix is $\boldsymbol{\Omega} = \boldsymbol{\Sigma}^{-1}$ and its information vector $\boldsymbol{\xi} = \boldsymbol{\Sigma}^{-1}\mu$. Algorithm 4.1 shows the Information Filter for updating the belief state $bel(\mathbf{x}_t) = \mathcal{N}(\boldsymbol{\Omega}_t^{-1}\boldsymbol{\xi}_t, \boldsymbol{\Omega}_t^{-1})$.

This representation has its advantages and drawbacks. One important advantage is that complete uncertainty is easily represented by $\boldsymbol{\Omega} = \mathbf{0}$.

A second interesting property is that the corresponding information matrix for the full trajectory is block-diagonal, and the space needed to store it is then linear with the trajectory length. Algorithm 4.2 shows the filter for the full belief trajectory. The functions **Augment_Matrix** and **Augment_Vector** augment the state space. However, as line 1 shows, the information matrix for the trajectory $\boldsymbol{\Omega}^t$ is block diagonal at any time (see Fig. 4.16).

Algorithm 4.1. $(\boldsymbol{\xi}_t, \boldsymbol{\Omega}_t) \leftarrow$ Information Filter$(\boldsymbol{\xi}_{t-1}, \boldsymbol{\Omega}_{t-1}, \mathbf{z}_t)$

1: $\boldsymbol{\Psi}_t = \mathbf{A}_t^{-T} \boldsymbol{\Omega}_{t-1} \mathbf{A}_t^{-1}$
2: $\bar{\boldsymbol{\Omega}}_t = \boldsymbol{\Psi}_t - \boldsymbol{\Psi}_t (\mathbf{R}_t^{-1} + \boldsymbol{\Psi}_t)^{-1} \boldsymbol{\Psi}_t$
3: $\bar{\boldsymbol{\xi}}_t = \bar{\boldsymbol{\Omega}}_t \mathbf{A}_t \boldsymbol{\Omega}_{t-1}^{-1} \boldsymbol{\xi}_{t-1}$
4: $p(\mathbf{x}_t | \mathbf{z}^{t-1}) \sim \mathcal{N}(\bar{\boldsymbol{\Omega}}_t^{-1} \bar{\boldsymbol{\xi}}_t, \bar{\boldsymbol{\Omega}}_t^{-1})$
5: $\boldsymbol{\Omega}_t = \bar{\boldsymbol{\Omega}}_t + \mathbf{M}_t^T \mathbf{S}_t^{-1} \mathbf{M}_t$
6: $\boldsymbol{\xi}_t = \bar{\boldsymbol{\xi}}_t + \mathbf{M}_t^T \mathbf{S}_t^{-1} \mathbf{z}_t$
7: $p(\mathbf{x}_t | \mathbf{z}^t) \sim \mathcal{N}(\boldsymbol{\Omega}_t^{-1} \boldsymbol{\xi}_t, \boldsymbol{\Omega}_t^{-1})$

Algorithm 4.2. $(\boldsymbol{\xi}^t, \boldsymbol{\Omega}^t) \leftarrow$Information Filter Trajectory$(\boldsymbol{\xi}^{t-1}, \boldsymbol{\Omega}^{t-1}, \mathbf{z}_t)$

1: $\bar{\boldsymbol{\Omega}}^t = \textbf{Augment\_Matrix}(\,\boldsymbol{\Omega}^{t-1}) + \begin{pmatrix} \begin{pmatrix} \mathbf{I} \\ -\mathbf{A}_t^T \end{pmatrix} \mathbf{R}_t^{-1} \left(\mathbf{I} \; -\mathbf{A}_t\right) & \mathbf{0}^T \\ \mathbf{0} & \mathbf{0} \end{pmatrix}$

2: $\bar{\boldsymbol{\xi}}^t = \textbf{Augment\_Vector}\;(\boldsymbol{\xi}^{t-1})$

3: $\boldsymbol{\Omega}^t = \bar{\boldsymbol{\Omega}}^t + \begin{pmatrix} \mathbf{M}_t^T \mathbf{S}_t^{-1} \mathbf{M}_t & \mathbf{0}^T \\ \mathbf{0} & \mathbf{0} \end{pmatrix}$

4: $\boldsymbol{\xi}^t = \bar{\boldsymbol{\xi}}_t + \begin{pmatrix} \mathbf{M}_t^T \mathbf{S}_t^{-1} \mathbf{z}_t \\ \mathbf{0} \end{pmatrix}$

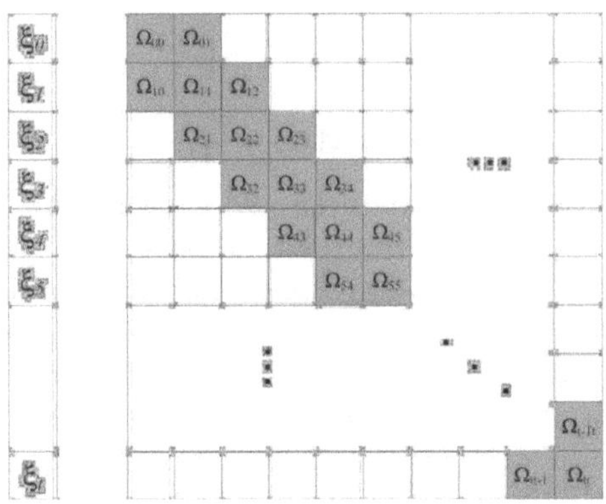

Fig. 4.16. Structure of the information matrix for the full trajectory. The information matrix is a block tridiagonal symmetric matrix.

4.5.1 Multi-robot Perception Employing Information Filters

It is straightforward to determine the Information Filter (IF) for the centralized multi-robot perception case. Recalling (4.7), the likelihood is given by $\prod_{j=1}^{M} p(\mathbf{z}_{j,t}|\mathbf{x}_t)$. If the assumptions of the IF hold, and as the local likelihood functions $p(\mathbf{z}_{j,t}|\mathbf{x}_t)$ are Gaussian, then this updating step consists of a sum of all the information contributions from all the robots of the fleet:

$$\boldsymbol{\Omega}_t = \bar{\boldsymbol{\Omega}}_t + \sum_{j=1}^{M} \mathbf{M}_{j,t}^T \mathbf{S}_{j,t}^{-1} \mathbf{M}_{j,t} \tag{4.35}$$

$$\boldsymbol{\xi}_t = \bar{\boldsymbol{\xi}}_t + \sum_{j=1}^{M} \mathbf{M}_{j,t}^T \mathbf{S}_{j,t}^{-1} \mathbf{z}_{j,t} \tag{4.36}$$

The centralized solution requires that the central node should receive all the information gathered by the UAVs. Even having a preprocessing stage carried on board the different UAVs, so that the central node does not receive the raw data, but only estimated features, the system does not scale well with the number of robots. Moreover, in this case the central node should be within communication range at any time. However, the IF leads to a decentralized implementation in which each UAV maintains a local belief employing only local information, and then shares this belief with the local neighbors.

In this case, each UAV will run locally Algorithm 4.1 (with $\mathbf{A}_t = \mathbf{I}$ and $\mathbf{R}_t = \mathbf{0}$ if the state is static) or 4.2. And, for updating the status of its alarms, it will use the information obtained from its sensors. Of course, the estimated belief state will be different from the centralized one due to the missed terms in the updating steps 5 and 6 (the information collected by the rest of the robots).

When UAV i flies within communication range with other UAV j, they can share their beliefs. UAV i can apply the relation (4.15) to update its belief state. Denoting by $bel_{i,m}(\mathbf{x}_t)$ the resultant belief state of UAV i after incorporating the belief from UAV j, and employing the logarithmic form of the combination rule (4.16):

$$
\begin{aligned}
bel_{i,m}(\mathbf{x}_t) = \log(C) + bel_i(\mathbf{x}_t) + bel_j(\mathbf{x}_t) - bel_j(\mathbf{x}_{t-1}) = \\
= \log(C) - \frac{1}{2}\mathbf{x}^T\boldsymbol{\Omega}_t^i\mathbf{x} + \mathbf{x}^T\boldsymbol{\xi}_t^i - \frac{1}{2}\mathbf{x}^T\boldsymbol{\Omega}_t^j\mathbf{x} + \mathbf{x}^T\boldsymbol{\xi}_t^j - \frac{1}{2}\mathbf{x}^T\boldsymbol{\Omega}_{t-1}^j\mathbf{x} + \mathbf{x}^T\boldsymbol{\xi}_{t-1}^j = \\
= \log(C) - \frac{1}{2}\left[\mathbf{x}^T(\boldsymbol{\Omega}_t^i + \boldsymbol{\Omega}_t^j - \boldsymbol{\Omega}_{t-1}^j)\mathbf{x}\right] + \mathbf{x}^T(\boldsymbol{\xi}_t^i + \boldsymbol{\xi}_t^j - \boldsymbol{\xi}_{t-1}^j)
\end{aligned}
\tag{4.37}
$$

Comparing this with the logarithmic of the Gaussian distribution it can be seen that the resultant information vector and matrix are updated with the increase in evidence $\boldsymbol{\Omega}_t^{i,m} = \boldsymbol{\Omega}_t^i + (\boldsymbol{\Omega}_t^j - \boldsymbol{\Omega}_{t-1}^j)$ and $\boldsymbol{\xi}_t^{i,m} = \boldsymbol{\xi}_t^i + (\boldsymbol{\xi}_t^j - \boldsymbol{\xi}_{t-1}^j)$, where $\boldsymbol{\Omega}_{t-1}^j, \boldsymbol{\xi}_{t-1}^j$ in this case denotes the last received belief from UAV j (whenever it was received).

If there are more than one UAV within communication range, then, if $C(i)$ is the set of neighbors of UAV i:

$$
\boldsymbol{\Omega}_t^{i,m} = \boldsymbol{\Omega}_t^i + \sum_{j \in C(i)} (\boldsymbol{\Omega}_t^j - \boldsymbol{\Omega}_{t-1}^j) \tag{4.38}
$$

$$
\boldsymbol{\xi}_t^{i,m} = \boldsymbol{\xi}_t^i + \sum_{j \in C(i)} (\boldsymbol{\xi}_t^j - \boldsymbol{\xi}_{t-1}^j) \tag{4.39}
$$

If the state is static the amount of information needed to store the belief state (and to communicate it) is constant along time. Moreover, one UAV does not have to communicate continuously its belief, as it can accumulate evidence and transmit it later (without increasing the storage). Therefore, from the point of view of bandwidth requirements, it could be adjusted depending on the network conditions.

If the state is dynamic, the UAVs update and transmit their trajectory estimations. This estimation grows with time, although only linearly. Nevertheless, the amount of information to be transmitted should be bounded. Each UAV only stores a trajectory interval (the last 20 seconds). The rest of the trajectory is marginalized out, which maintains the sparse structure of the information matrix.

Eliminating Common Information

The previous equations, as commented in Sect. 4.2.3, assume that there are no loops in the network of UAVs, in the sense that the belief information shared between two robots follows an unique path. If it is not the case, prior to combining the beliefs, the common information should be removed.

There are several options that can be considered for avoiding an overconfident estimation due to accounting common information several times. The first one is to ensure a tree topology in the belief network, as seen in Sect. 4.2.3. However, this option imposes a strong constraint on the potential communication links among the UAVs.

Another option is that each UAV only sends its last local information update at every time instant. This way, no information is duplicated. The problem in this case is that the use of an UAV as *data mule* is lost: one UAV that collects the evidence from a group of local neighbors will communicate it to other robots that could be initially disconnected from the firsts. Moreover, if the connection between two UAVs is lost, it will lose information that would have been available in the case that the robot had sent the complete belief.

The last option is to employ a conservative fusion rule, which ensures that the robot does not become overconfident. For the case of the information filter, there is an analytic solution for this, given by the *Covariance Intersection* (CI) algorithm [22]. The CI algorithm is a way of combining information from random variables whose cross-correlations are unknown.

Therefore, the conservative rule to combine the belief of UAV i with that received from the set $C(i)$ of local neighbors is given by:

$$\boldsymbol{\Omega}_t^{i,m} = \omega \boldsymbol{\Omega}_t^i + (1-\omega) \sum_{j \in C(i)} \boldsymbol{\Omega}_t^j \tag{4.40}$$

$$\boldsymbol{\xi}_t^{i,m} = \omega \boldsymbol{\xi}_t^i + (1-\omega) \sum_{j \in C(i)} \boldsymbol{\xi}_t^j \tag{4.41}$$

for $\omega \in [0\ 1]$. It can be seen that the estimation is consistent in the sense that $\boldsymbol{\Sigma}^{i,m} - \hat{\boldsymbol{\Sigma}}^i \geq \mathbf{0}$ (where $\boldsymbol{\Sigma}^{i,m} = \boldsymbol{\Omega}^{i,m-1}$ is the estimated covariance matrix and $\hat{\boldsymbol{\Sigma}}^i$ is the actual covariance matrix) for any (unknown) cross-correlation matrix $\boldsymbol{\Sigma}^{ij}$, and for any ω. The value of ω can be selected following some criteria, as maximizing the obtained determinant of $\boldsymbol{\Omega}^{i,m}$ (minimizing the entropy of the final distribution). Another option is to use it as a weight that shows the UAV confidence in its own estimation and the neighbor's ones.

4.5.2 Decentralized Information Filter for Object Detection and Localization

Recalling what was presented in Sect. 4.3, the objective is to estimate the state of the potential alarms present in an scenario. In the general case that the objects move, the state will be the position and velocity of all the alarms, and its nature.

$$\mathbf{x}_t = [\mathbf{p}_{1,t}^T, \dot{\mathbf{p}}_{1,t}^T, \boldsymbol{\theta}_1, \ldots, \mathbf{p}_{N_t,t}^T, \dot{\mathbf{p}}_{N_t,t}^T, \boldsymbol{\theta}_{N_t}]^T \tag{4.42}$$

Each UAV will update its knowledge locally employing the Information Filter. The local estimations will be shared with the other UAVs of the fleet, which will combine them by using the relations given by (4.40) and (4.41) in order to avoid overconfident estimates in the fully decentralized case.

Motion Model

In general, the motion of the alarms will not depend on the others. That is, at time t, $\mathbf{A}_t = diag\{\mathbf{A}_{1,t}, \ldots, \mathbf{A}_{N_t,t}\}$. Also $\mathbf{R}_t = diag\{\mathbf{R}_{1,t}, \ldots, \mathbf{R}_{N_t,t}\}$.

As a generic motion model of the targets, a discrete version of the continuous white noise acceleration model or second-order kinematic model is used [42]. In this model, the velocity is assumed to be affected by an acceleration modeled as a white noise of zero mean and with power spectral density $\boldsymbol{\delta}$. The discretized version of this linear motion model for an object i, is characterized by:

$$\mathbf{A}_{i,t} = \begin{pmatrix} \mathbf{I} & \Delta t\mathbf{I} \\ 0 & \mathbf{I} \end{pmatrix} \tag{4.43}$$

and

$$\mathbf{R}_{i,t} = \begin{pmatrix} \frac{1}{3}\Delta t^3\mathbf{I} & \frac{1}{2}\Delta t^2\mathbf{I} \\ \frac{1}{2}\Delta t^2\mathbf{I} & \Delta t\mathbf{I} \end{pmatrix} \boldsymbol{\delta} \tag{4.44}$$

Measurement Model and Likelihood Function

If the alarms are located on the ground, and if the geolocation procedure described above can be applied, each UAV can determine directly the 3D position of the objects segmented on the image plane. If a measurement j is associated to a particular alarm i, then:

$$\mathbf{M}_{ji,t} = \begin{pmatrix} \mathbf{I}_i & \mathbf{0} \end{pmatrix} \tag{4.45}$$

The errors $\boldsymbol{\varepsilon}_{j,t}$ in the estimated position arise due to the errors on the position and orientation of the sensor ($\mathbf{q}_t$) and the terrain model. Moreover, the geolocation procedure is non-linear. In order to propagate the uncertainties in $\mathbf{q}_t$ and obtain an estimation of the covariances of the error in (4.34) the Unscented Transform (UT) is used. Therefore, the UT is used to determine the mean $\mathbf{z}_{j,t}$ and covariance matrix $\mathbf{S}_{j,t}$ on the estimated position for any alarm detected on the image plane.

Prior Belief

The only issue to be described is how alarms are initialized. When an UAV segments an alarm, the geolocation procedure directly provides the initial values for $\boldsymbol{\xi}_i$ and $\boldsymbol{\Omega}_i$:

$$\boldsymbol{\Omega}_i = \begin{pmatrix} \mathbf{S}_{j,t}^{-1} & \mathbf{0} \\ \mathbf{0} & \mathbf{0} \end{pmatrix} \tag{4.46}$$

and

$$\boldsymbol{\xi}_i = \begin{pmatrix} \mathbf{S}_{j,t}^{-1}\mathbf{z}_{j,t} \\ \mathbf{0} \end{pmatrix} \tag{4.47}$$

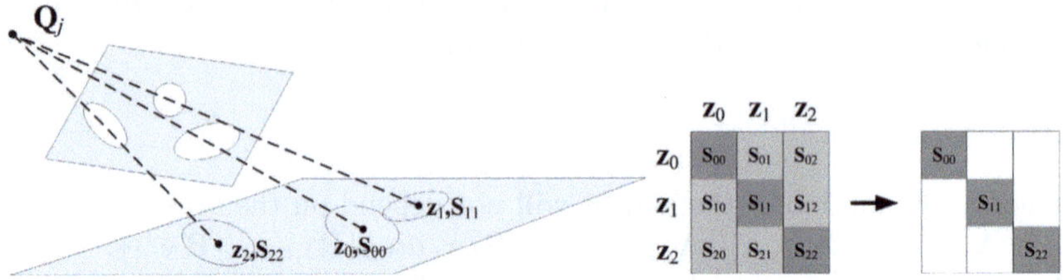

Fig. 4.17. The errors in $\mathbf{q}_t$ induce correlated errors in the estimated position of the alarms $\mathbf{z}_{i,t}$ (left). If the alarms are to be evolved independently, the dependences should be marginalized out (right).

Local Information Filter with Perfect Data Association

The previous equations are then used locally by each UAV to update their local beliefs about the current alarms. A further assumption allows to simplify the algorithm. As the motion model and motion noise $\boldsymbol{\nu}_t$ covariances are block diagonal, if the information matrix is block diagonal $\boldsymbol{\Omega}_{t-1} = diag\{\boldsymbol{\Omega}_{1,t-1}, \ldots, \boldsymbol{\Omega}_{N_t,t-1}\}$, each alarm i can be predicted separately leading to a (local) parallelized computation.

If the set of measurements can be associated with the current set of alarms, so that any measurement j is associated to an alarm i (that is, only one block $\mathbf{M}_{ji,t}$ for each row of matrix $\mathbf{M}_t$ is non zero) and if the measurements are independent (and thus, $\mathbf{Q}_t$ is block diagonal), the global Information Filter for all the alarms can be divided into N Information Filters, a separated one for each alarm.

However, the position measurements about the different alarms obtained by one UAV are not independent. When obtaining the position of the alarms, the errors on the UAV position $\mathbf{q}_t$ induce errors on the estimated position, errors that are correlated for all the alarms detected (see Fig. 4.17). Therefore, the dependencies among measurements should be explicitly marginalized out if one wants to keep the alarms updated independently. The marginalization is straightforward for the case of Gaussian measurements. It should be noted that getting rid of measurements dependencies reduce the amount of storage required and the computational burden, but it comes at a cost. Some information is lost: the relation between alarms. This information would allow to propagate information about one alarm to the other related (as in the SLAM problem).

Data Association

Previous sections have presented the filter under the assumption of perfect data association. That is, each measurement j is associated to one particular alarm i, or when an UAV receives information from other, it knows to which alarm it belongs.

The data association problem tries to determine what measurements correspond to what alarms, or what alarm corresponds to what alarm when combining beliefs received from other UAVs. The first one is usually called scan-to-track association, while the later is called track-to-track association. This general problem is known to be NP-hard in the general case. As the main objective is to show cooperative characteristics, the approach followed is fairly simple, although more complex ones could be used.

For the scan-to-track association a gated nearest neighbor technique [10] is employed. For the track-to-track association, each alarm is marked with a label, but the labels are not synchronized, so the UAVs must determine the conversion. This is also accomplished by using a nearest neighbor technique, so that two alarms p and q of UAVs i and j are associated if the Mahalanobis distance is below a threshold:

$$d_{kp}^{ij2} = [\boldsymbol{\mu}_k^i - \boldsymbol{\mu}_p^j]^T \boldsymbol{\Sigma}_{kp}^{ij-1} [\boldsymbol{\mu}_k^i - \boldsymbol{\mu}_p^j] \leq d_{\text{th}}^2 \tag{4.48}$$

In the case of combining belief trajectories, it is needed to recover the mean $\boldsymbol{\mu}^i$ for every alarm by solving the system $\boldsymbol{\Omega}^i \boldsymbol{\mu}^i = \boldsymbol{\xi}^i$. Also, it is needed to obtain the inverse of $\boldsymbol{\Sigma}^{ij}$. The information matrix $\boldsymbol{\Omega}^i$ for the case of the full trajectory (and, thus, matrix $\boldsymbol{\Sigma}^{ij}$) can be of the order of hundreds of rows/columns (for instance, for a 20 seconds trajectory and if each block row corresponds to one second, the matrix is 120×120). However, the information matrix is very structured, which allows for efficient algorithms for matrix inversion. In [2], the authors show how for a symmetric tridiagonal block matrix there exist algorithms that are nearly two orders of magnitude more efficient than direct inversion.

In the case of track-to-track association, once the same identifier has been associated to a local alarm the track is definitely associated, so that the data association problem becomes straightforward. However, this comes at cost that sometimes this could lead to wrong assignments. More complex data association techniques can be used. An approach based on the information representation of the posterior over all potential track-to-track associations is described in [37].

4.5.3 Results

This section presents some results obtained in simulation. The same algorithms have been applied for real experiments of fire detection and monitoring, which are presented in Chap. 8.

Experiment 1

In the first experiments, three vehicles are considered, and two static alarms. Figure 4.18 shows the simulated trajectories and the position of the alarms. One of the vehicles flies at a higher altitude, while the others are given a certain area to patrol. The vehicle flying at higher altitude acts as an information relay for the other two UAVs.

Figures 4.19 and 4.20 show the estimated positions and estimated variances for the two objects and 3 UAVs. The UAVs employ the Covariance Intersection to avoid overconfident estimates. It can be seen as all converge to the same solution and to the correct position of the alarms.

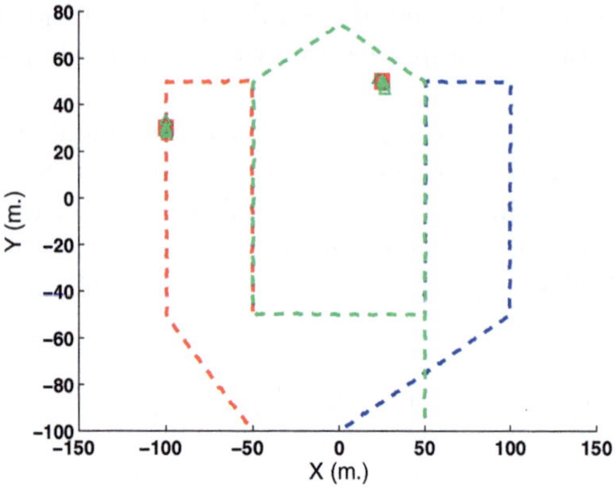

Fig. 4.18. Map showing the trajectories of the vehicles and the position of the events for Experiment 1

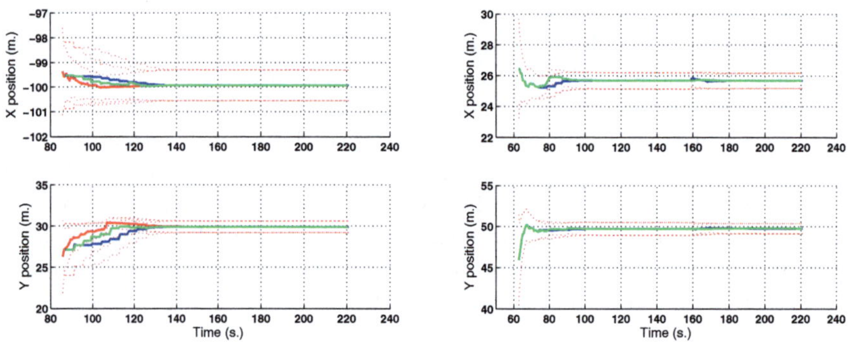

Fig. 4.19. Estimated position of both alarms for the three vehicles in Experiment 1

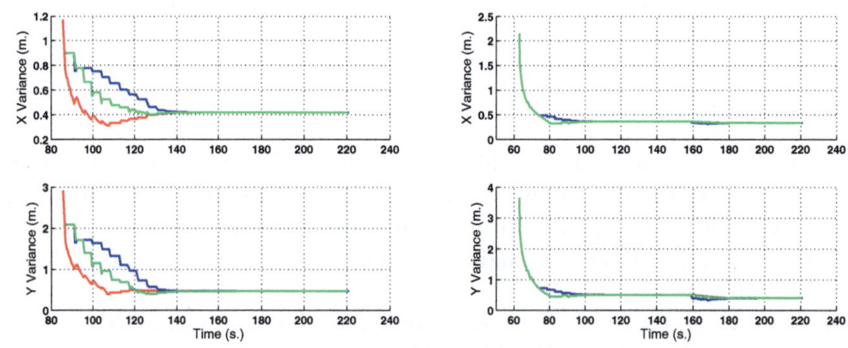

Fig. 4.20. Estimated variances of the errors for the three vehicles in Experiment 1

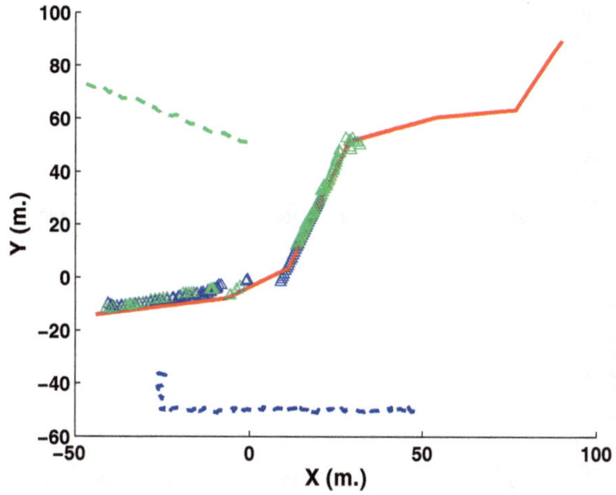

Fig. 4.21. Map showing the trajectory of the event (solid), the trajectories of the two vehicles (dashed) and the estimated position of the object (triangles) in Experiment 2

Experiment 2

The second experiment consists of the detection and tracking of a moving object. Two UAVs are commanded to fly to a certain point and hover looking to the object. Figure 4.21 shows the estimated position of a moving object by two UAVs using cameras.

Figure 4.22 shows a detail of the estimated position by the UAV located around position $(0, 50)$ against the actual one. It can be seen that first the object is within its field of view. One it abandons it, the uncertainty grows until

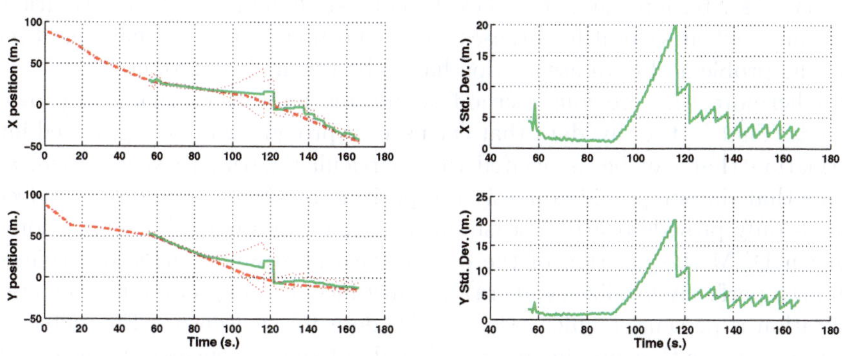

Fig. 4.22. Left: Estimated position of the object (solid) and actual one (dashed). Right: estimated variances for one of the UAVs. At time 115 it receives information from the other UAV, incorporating it into its own belief.

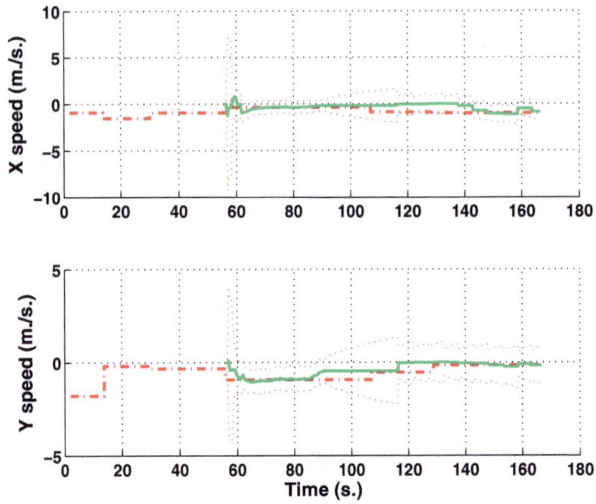

Fig. 4.23. Estimated velocity of the object (solid) and actual one (dashed)

it receives information from the other UAV. Also, the variances in the X and Y coordinates are shown. Figure 4.23 shows the estimated velocity of the object for the same UAV.

4.6 Grid-Based Multi-UAV Detection and Localization Using Vision and Other Sensors

The Information Filter (and its dual, the Kalman Filter), requires some restrictions that in some cases do not hold. For instance, they are restricted to Gaussian distributions, and thus they cannot handle multi-modal distributions, which arise when several hypotheses should be maintained at the same time. Also, the problem of data association has to be taken into account [42].

Furthermore, the previous method is suitable for sensors that can provide *contacts*. That is, the raw data that each sensor provides can be segmented into information that can be associated to a particular alarm. However, there are sensors that do not provide directly the position of objects of interest (indeed, cameras only provide bearing measurements, and only under the assumption of a known DEM the position of the objects on the ground can be determined, even for a perfectly localized camera). For instance, in Chapter 8 a fire detection application is presented which also uses fire detectors as sensors. Fire detectors are cheap sensors that provide information about the presence or absence of fire within their fields of view, but no direct information about the actual localization or size of the fire.

Grid-based approaches [27, 41, 46] can overcome the previous mentioned problems. The approach is to divide the scenario to be explored into cells, in what is called a certainty or evidence grid. To each cell k, a discrete (binary) random

Algorithm 4.3. $\{l_{k,t}\} \leftarrow$ Binary_LogOdds_Filter($\{l_{k,t-1}\}, \mathbf{z}_t$)

1: **for** $k = 1$ to L **do**
2: $\quad l_{k,t} = l_{k,t-1} + \log p(\mathbf{z}_{i,t}|x_k) - \log p(\mathbf{z}_{i,t}|\bar{x}_k)$
3: **end for**

variable x_k is attached, representing the presence or absence of the object. Also, each cell of the grid has a 3D position associated, $\mathbf{p}_k$.

The variable can have two values, true or false. The probability that there is an object at cell k is represented by $p(x_k = 1)$ or $p(x_k)$, and we denote by $\bar{x}_k$ the fact that there is no object at cell k. Then, by definition, $p(x_k = 0) = p(\bar{x}_k) = 1 - p(x_k)$. The objective of the robot team is to update the probability of the different cells.

4.6.1 Local Filter

For this application, it is assumed that the status of each cell does not change with time. The state $\mathbf{x}_t$ is comprised by the status of all the cells of the grid $x_{k,t}$ at time t. The joint posterior $p(\mathbf{x}_t|\mathbf{z}^t)$ has to take into account all the possible combinations of values for all the cells. The full posterior for a grid with L cells should consider 2^L different states [46]. Maintaining this posterior is computationally unaffordable. Instead, this posterior will be approximated by the products of its marginals over each cell.

$$p(\mathbf{x}_t|\mathbf{z}^t) = \prod_k p(x_k|\mathbf{z}^t) \tag{4.49}$$

In this case, the Bayes filter takes a particular simple form considering log-odds:

$$\frac{bel(x_{k,t})}{1 - bel(x_{k,t})} = \frac{p(\mathbf{z}_t|x_k)}{p(\mathbf{z}_t|\bar{x}_k)} \frac{bel(x_{k,t-1})}{1 - bel(x_{k,t-1})} = = \prod_{\tau=0}^{t} \frac{p(\mathbf{z}_\tau|x_k)}{p(\mathbf{z}_\tau|\bar{x}_k)} \frac{bel(x_{k,0})}{1 - bel(x_{k,0})} \tag{4.50}$$

Calling $l_{k,t} = \log \frac{bel(x_{k,t})}{1-bel(x_{k,t})}$, then the local algorithm for updating all the cells of the grid is given by Algorithm 4.3.

4.6.2 Distributed Filter in the Multi-robot Case

If a centralized node receives all the data provided by the robots, as $p(\mathbf{z}_t^m|x_{k,t}) = \prod_j p(\mathbf{z}_{j,t}|x_{k,t})$ the central filter is the same as Algorithm 4.3, but line 2 is substituted by

$$l_{k,t} = l_{k,t-1} + \sum_j [\log p(\mathbf{z}_{j,t}|x_k) - \log p(\mathbf{z}_{j,t}|\bar{x}_k)] \tag{4.51}$$

The filter is easily decentralized. Each robot i computes part of the running total of (4.51) considering only local data, and sends to the other robots its own belief state in logarithmic form. Each robot incorporates the information received from others by using (4.37) adapted to the binary static case. The final filter is given by Algorithm 4.4.

Algorithm 4.4. Decentralized_Grid_Multi_Robot(i)

for all k **do**
  $l^i_{k,0} = 0$
end for
while true **do**
  for all k **do**
    if New data $\mathbf{z}_{i,t}$ **then**
      $l^i_{k,t} \leftarrow l^i_{k,t-1} + \log p(\mathbf{z}_{i,t}|x_k) - \log p(\mathbf{z}_{i,t}|\bar{x}_k)$
    end if
    if New belief from UAV j **then**
      $l^i_{k,t} \leftarrow l^i_{k,t} + l^j_{k,t} - l^j_{k,t-\Delta t}$
    end if
  end for
end while

4.6.3 Grid-Based Detection and Localization of Events

The main issue in the filter described is to determine the likelihood function $p(\mathbf{z}_t|x_k)$. It indicates the probability of having data $\mathbf{z}_t$ considering that there is an object of the class considered in cell k at time t. The data $\mathbf{z}_t$ consist of all the data gathered by the vehicles of the fleet at time t, and in the particular case considered, these data are images and fire sensor readings gathered by the different vehicles.

The likelihood function should take into account the position of the sensors respect to the map and the geometric characteristics of the distinct sensors (for instance the pin-hole model of the cameras). The latter are obtained through calibration, while the former will be provided by the UAVs. As commented above, the uncertainty on the pose of each UAV j, $\mathbf{q}_{j,t}$, should be taken into account for a correct definition of the likelihood function for the measurements gathered by this UAV $\mathbf{z}_{j,t}$:

$$p(\mathbf{z}_{j,t}|x_k) = \int p(\mathbf{z}_{j,t}|x_k, \mathbf{q}_{j,t})p(\mathbf{q}_{j,t})d\mathbf{q}_{j,t} \tag{4.52}$$

An equivalent equation to (4.52) is used to compute $p(\mathbf{z}_{j,t}|\bar{x}_k)$ (which is required in the filter). Equation (4.52) implies a further simplification. The measurements depend conditionally not only in one cell, but at least in all the cells of the grid within their field of view $S(j)$ of the sensor j. Therefore,

$$p(\mathbf{z}_{j,t}|x_k, \mathbf{q}_{j,t}) = \sum_{i \in S(j)} \sum_{x_i} p(\mathbf{z}_{j,t}|x_k, x_i, \mathbf{q}_{j,t})p(x_i) \tag{4.53}$$

However, when computing (4.52) this dependence will not be considered. The rest of the section describes the likelihood functions for several sensors.

Measurement Model for the Cameras

As presented in Sect. 4.3.3, cameras are an important source of information in the fleet of UAVs. Cameras of different modalities can be considered, but the first assumption is that in a preprocessing stage, objects of interest are segmented from the background, so that the measurements provided by the cameras are binary images that classify the pixels as belonging to the class or not (as fire or no fire for instance). This segmentation algorithms are defined by the probabilities $P_{\mathrm{D},j}$ and $P_{\mathrm{F},j}$ for sensor j.

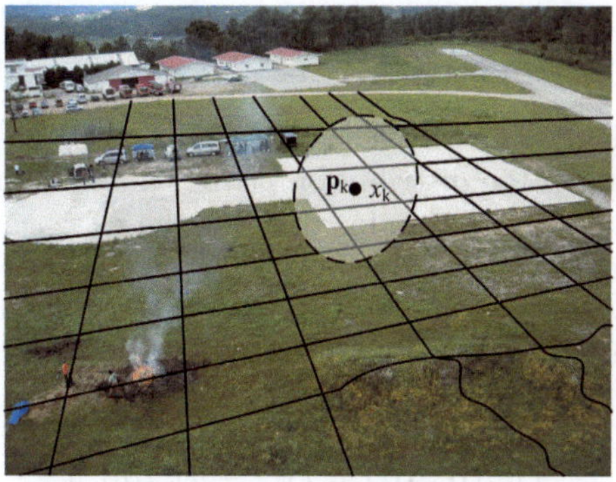

Fig. 4.24. Due to the uncertainties in sensor position and the cell resolution, one cell corresponds to a zone on the image plane (represented by the ellipse)

In order to determine the likelihood function (4.52), it should be considered the following. Each cell k has a position associated, $\mathbf{p}_k$. For a given value of the position of the sensor $\mathbf{q}_{j,t}$, the center of the cell will correspond to a pixel $\mathbf{m}_{k,j}$ on the image plane of camera j (if it is within the field of view of that camera). The pixel position is given by (4.25).

If pixel $\mathbf{m}_{k,j}$ corresponds to a region segmented as fire, then the likelihood is defined by $p(\mathbf{z}_{j,t}|x_k, \mathbf{q}_{j,t})$:

$$\begin{aligned} p(\mathbf{z}_{j,t}|x_k, \mathbf{q}_{j,t}) &= P_{\mathrm{D},j} \\ p(\mathbf{z}_{j,t}|\bar{x}_k, \mathbf{q}_{j,t}) &= P_{\mathrm{F},j} \end{aligned} \tag{4.54}$$

while if the pixel is classified as background, then the term is given by:

$$\begin{aligned} p(\mathbf{z}_{j,t}|x_k, \mathbf{q}_{j,t}) &= 1 - P_{\mathrm{D},j} \\ p(\mathbf{z}_{j,t}|\bar{x}_k, \mathbf{q}_{j,t}) &= 1 - P_{\mathrm{F},j} \end{aligned} \tag{4.55}$$

However, the position of the sensor is not known accurately, and thus, the position of the corresponding pixel is also uncertain. So in order to compute the likelihood $p(\mathbf{z}_{j,t}|x_k)$, (4.52) should be integrated for possible values of the pose of the camera $\mathbf{q}_{j,t}$. This could be done by sampling values of $\mathbf{q}_{j,t}$. However, this is computationally expensive, and indeed this should be done for all the cells of the grid that are within the field of view of the camera.

The pixel position is related to the sensor and cell position through the non-linear pin-hole model of (4.25), so that $\mathbf{m}_{kj} = \mathbf{f}(\mathbf{q}_{j,t}, \mathbf{p}_k)$. Instead of directly solving (4.52), the uncertainties in $\mathbf{q}_{j,t}$ are propagated into uncertainties on the pixel position $\mathbf{m}_{kj}$ corresponding to cell k using the Unscented Transform [21]. Moreover, in the procedure, the uncertainties in the position $\mathbf{p}_k$ due to the limited resolution of the grid are also taken into account. As a result, each cell k corresponds to a Gaussian distribution on the pixel position $p(\mathbf{m}_{kj})$. And then, equation (4.52) becomes:

$$p(\mathbf{z}_{j,t}|x_k) = \int p(\mathbf{z}_{j,t}|x_k, \mathbf{q}_{j,t})p(\mathbf{q}_{j,t})\mathrm{d}\mathbf{q}_{j,t} = \sum_{\mathbf{m}} p(\mathbf{z}_{j,t}|\mathbf{m}_{kj,t})p(\mathbf{m}_{kj,t}) \quad (4.56)$$

where the sum is done over a region on the image plane determined by the second order moments of the distribution $p(\mathbf{m}_{jk,t})$. The same procedure is used to compute the likelihood function for the hypotheses $\bar{x}_k$.

To complete the model, the probabilities $P_{\mathrm{D},j}$ and $P_{\mathrm{F},j}$ are modified depending mainly on the relative position of the cell k, $\mathbf{p}_k$, respect to the position and orientation of the sensor $\mathbf{q}_j$. Thus:

$$\begin{aligned} P_{\mathrm{D},j}(\mathbf{p}_k, \mathbf{q}_j) &= P_{\mathrm{D},j} - w_{\mathrm{D},j}(d_{kj}^2) \\ P_{\mathrm{F},j}(\mathbf{p}_k, \mathbf{q}_j) &= P_{\mathrm{F},j} - w_{\mathrm{F},j}(d_{kj}^2) \end{aligned} \quad (4.57)$$

where $w_{\mathrm{D},j}$ and $w_{\mathrm{F},j}$ are functions that decrease the values of $P_{\mathrm{D},j}$ and $P_{\mathrm{F},j}$ with the distance between cell k and sensor j, d_{kj}. These are the actual values employed by the likelihood function.

Measurement Model for the Fire Sensor

The fire sensor considered is a fire detector, whose main component is a photodiode set-up to limit its sensibility to the band of [185, 260] nm, normally associated to fires. The output of the sensor is a scalar value, proportional to the radiation received. Being a magnitude sensor, it is not possible to determine if a measure is due to a big fire far away or a nearby small fire. Using a threshold, this value is used to indicate if a fire is present or not within the field of view of the sensor.

Thus, the functioning of the sensor can be also characterized by probabilities of detection P_{D} and false positive generation P_{F}. These probabilities depend on the threshold selected. A higher threshold implies a lower P_{F} at a cost of worse detection capabilities.

If the sensor detects something, then for the updating of the cells the following values hold:

$$\begin{aligned} p(\mathbf{z}_{j,t}|x_k) &= P_{\mathrm{D},j}(\mathbf{p}_k, \mathbf{q}_j) = P_{\mathrm{D},j} - w_{\mathrm{D},j}(d_{kj}^2, \alpha_{kj}, \theta_{kj}) \\ p(\mathbf{z}_{j,t}|\bar{x}_k) &= P_{\mathrm{F},j}(\mathbf{p}_k, \mathbf{q}_j) = P_{\mathrm{F},j} - w_{\mathrm{F},j}(d_{kj}^2, \alpha_{kj}, \theta_{kj}) \end{aligned} \quad (4.58)$$

and, in case that the sensor does not detect anything:

$$\begin{aligned} p(\mathbf{z}_{j,t}|x_k) &= 1 - P_{\mathrm{D},j}(\mathbf{p}_k, \mathbf{q}_j) = 1 - [P_{\mathrm{D},j} - w_{\mathrm{D},j}(d_{kj}^2, \alpha_{kj}, \theta_{kj})] \\ p(\mathbf{z}_{j,t}|\bar{x}_k) &= 1 - P_{\mathrm{F},j}(\mathbf{p}_k, \mathbf{q}_j) = 1 - [P_{\mathrm{F},j} - w_{\mathrm{F},j}(d_{kj}^2, \alpha_{kj}, \theta_{kj})] \end{aligned} \quad (4.59)$$

In fact, this model could be used for any other presence sensor that provides binary decisions about the presence or absence of an object of a certain kind within its field of view.

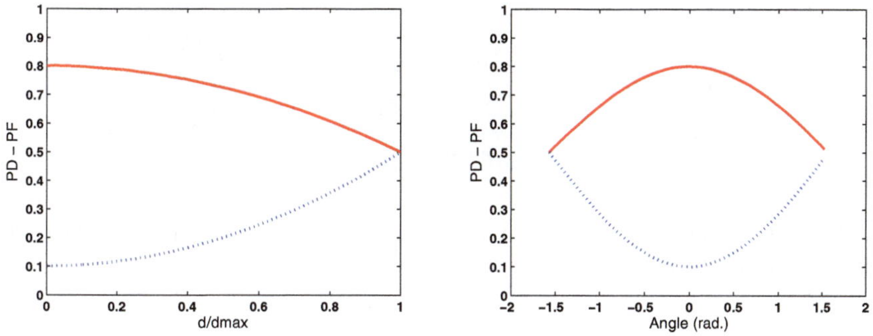

Fig. 4.25. Plots of equations $P_{\mathrm{D},j}(d^2_{kj}, \alpha_{kj}, \theta_{kj})$ and $P_{\mathrm{F},j}(d^2_{kj}, \alpha_{kj}, \theta_{kj})$. Left, distance component; right, angular component. Solid, P_{D}, dotted, P_{F}.

Obtaining Measures from the Grid

Using the equations described above, the status of the grid is recursively estimated using the data the vehicles are providing. From a Bayesian point of view, the grid represents all the information about the possible alarms at time t. However, in some applications, more specific measures are required. For instance, if a fleet is looking for fire alarms, a control center would expect the position of the potential fire alarm detected, in order to plan a new mission, sending new vehicles to confirm the alarm. Also, we will use this value to compare it with the position of the fire recorded with GPS for validation purposes.

This can be accomplished in various ways. In this case, the set of cells of the grid with probabilities over a given threshold is obtained every T seconds. An alarm is raised for each set R of connected cells over this threshold. The position of the alarm is computed as the weighted geometric mean of the positions of the cells.

$$\mu_R = \frac{\sum_{k \subset R} \mathbf{p}_k p(x_k|\mathbf{z}^t)}{\sum_{k \subset R} p(x_k|\mathbf{z}^t)} \tag{4.60}$$

Also, it can be obtained an estimation of the uncertainty on the computed position from the second order moments of the region R.

Experimental Results

Figure 4.26 shows the evolution of the grid of one UAV in several phases of an actual fire experiment carried out in the framework of the COMETS project, and that are described in Chap. 8. The first figure shows the status of the grid after one of the UAVs has flown over a place with no fire, using a fire detector. The second figure shows how the sensor produces two big high probability blobs on the grid, one due to a false alarm and other due to the actual alarm. Afterwards, another UAV takes off and uses its IR camera over the zone of the possible alarms. This UAV receives the estimated grid from the other one. The third grid shows how after several images and fire sensor data are integrated, the high

Fig. 4.26. The status of the gird at three moments during the mission. The filled square represents the actual position of the fire.

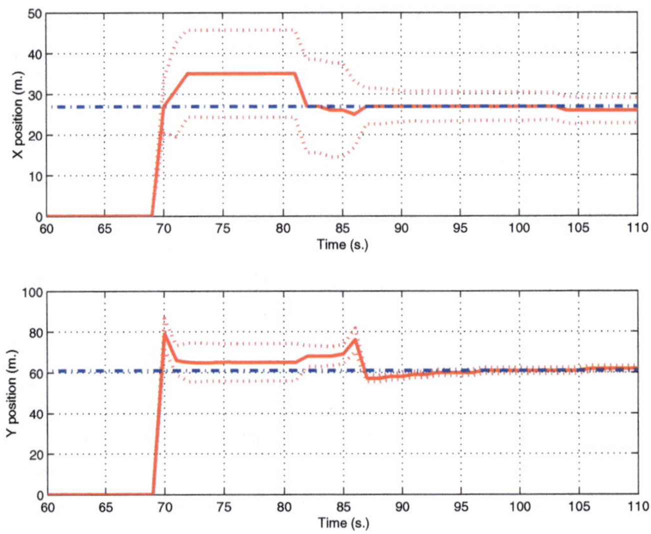

Fig. 4.27. Estimated mean position of one of the high probability regions. Dotted: estimated variances. Dash-dotted: actual fire position.

probability region is constrained to a smaller region, which includes the actual position of the fire.

Figure 4.27 shows the evolution of the position of the high probability regions computed using (4.60) compared to the actual fire position. It also shows the estimation on the uncertainty on the computed position.

4.7 Conclusions

UAV environment perception is a main issue in aerial robotics. UAV perception techniques include motion estimation from images, stabilization in sequences of images taken with on-board cameras subject to vibrations and UAV turbulences, automatic detection, classification and geo-localization of objects in the images, and UAV vision-based localization, which can be very useful in case of GPS unavailability or failures. Different methods for the application of these techniques have been presented in this Chapter.

The Chapter has also shown that a probabilistic approach is suitable for cooperative detection, localization and tracking. Particularly, an information filter for multi-UAV cooperative perception has been presented showing good results in detection and tracking. Moreover, a grid-based method for detection and tracking using vision and other sensors has been also presented demonstrating its capability for fire detection and tracking.

The presented methods can be applied integrated in the decisional architecture presented in Chap. 2. Chapter 8 of this book will present experiments on forest fire detection and monitoring.

References

1. Blobdetect software. `http://www.isy.liu.se/~perfo/software/`.
2. A. Asif and J.M.F. Moura. Block matrices with L-block-banded inverse: inversion algorithms. *IEEE Transactions on Signal Processing*, 53(2):630–642, Feb. 2005.
3. N. Ayache and P. T. Sander. *Artificial Vision for Mobile Robots: Stereo Vision and Multisensory Perception.* The MIT Press, Cambridge, MA, USA, 1991.
4. T. Balch and R.C. Arkin. Behavior-based formation control for multirobot teams. *IEEE Transactions on Robotics and Automation*, 14(6):926–939, 1998.
5. W. Burgard, M. Moors, C. Stachniss, and F.E. Schneider. Coordinated multi-robot exploration. *IEEE Transactions on Robotics*, 21(3):376–386, June 2005.
6. F. Caballero, L. Merino, J. Ferruz, and A. Ollero. A visual odometer without 3d reconstruction for aerial vehicles. applications to building inspection. In *Proceedings of the International Conference on Robotics and Automation*, pages 4684–4689. IEEE, April 2005.
7. D.W. Casbeer, D.B. Kingston, R.W. Bear, T.W. McLain, and S.M. Li. Cooperative forest fire surveillance using a team of small unmanned air vehicles. *International Journal of System Science*, pages 1–18, January 2005.
8. G. Farnebäck and K. Nordberg. Motion detection in the WITAS project. In *Proceedings SSAB02 Symposium on Image Analysis*, pages 99–102, Lund, March 2002. SSAB.
9. O. Faugeras. *Three-dimensional computer vision: a geometric viewpoint.* The MIT Press, Cambridge, MA, USA, 1993.
10. H. J. S. Feder, J. J. Leonard, and C. M. Smith. Adaptive Mobile Robot Navigation and Mapping. *The International Journal of Robotics Research*, 18(7):650–668, 1999.
11. J.W. Fenwick, P.M. Newman, and J.J. Leonard. Cooperative concurrent mapping and localization. In *Proceedings of the International Conference on Robotics and Automation*, pages 1810–1817, 2002.
12. J. Ferruz and A. Ollero. Real-time feature matching in image sequences for non-structured environments. applications to vehicle guidance. *Journal of Intelligent and Robotic Systems*, 28:85–123, 2000.
13. P.-E. Forssén and A. Moe. View matching with blob features. In *2nd Canadian Conference on Robot Vision*, pages 228–235, Victoria, BC, Canada, May 2005. IEEE Computer Society.
14. Per-Erik Forssén. *Low and Medium Level Vision using Channel Representations.* PhD thesis, Linköping University, 2004. Thesis No. 858.

15. Robert Grabowski, L.E. Navarro-Serment, C.J.J. Paredis, and P.K. Khosla. Heterogeneous teams of modular robots for mapping and exploration. *Autonomous Robots*, 8(3):293–308, 2000.
16. S. Grime and H. F. Durrant-Whyte. Data fusion in decentralized sensor networks. *Control Engineering Practice*, 2(5):849–863, Oct. 1994.
17. B. Grocholsky, A. Makarenko, and H. Durrant-Whyte. Information-theoretic coordinated control of multiple sensor platforms. In *Proceedings of the International Conference on Robotics and Automation*, pages 1521–1526, September 2003.
18. R. I. Hartley and A. Zisserman. *Multiple View Geometry in Computer Vision*. Cambridge University Press, second edition, 2004.
19. A. Howard, L.E. Parker, and G.S. Sukhatme. Experiments with a Large Heterogeneous Mobile Robot Team: Exploration, Mapping, Deployment and Detection. *The International Journal of Robotics Research*, 25(5-6):431–447, 2006.
20. E. Hygounenc, I-K. Jung, P. Soueres, and S. Lacroix. The Autonomous Blimp Project of LAAS-CNRS: Achievements in Flight Control and Terrain Mapping. *The International Journal of Robotics Research*, 23(4-5):473–511, 2004.
21. S. Julier and J. Uhlmann. A new extension of the Kalman filter to nonlinear systems. In *Proceedings of the 11th Int. Symp. on Aerospace/Defence Sensing, Simulation and Controls*, 1997.
22. S.J. Julier and J.K. Uhlmann. A non-divergent estimation algorithm in the presence of unknown correlations. In *Proceedings of the American Control Conference*, volume 4, pages 2369–2373, Jun. 1997.
23. Tony Lindeberg. *Scale-space Theory in Computer Vision.* Kluwer Academic Publishers, 1994. ISBN 0792394186.
24. M.J. Matarič. Minimizing complexity in controlling a mobile robot population. In *Proceedings of the International Conference on Robotics and Automation*, pages 830–835, 1992.
25. L. Merino, F. Caballero, J. R. Martinez, J. Ferruz, and A. Ollero. A cooperative perception system for multiple uavs: Application to automatic detection of forest fires. *Journal of Field Robotics*, 23(3):165–184, 2006.
26. L. Merino, J. Wiklund, F. Caballero, A. Moe, J.R. Martínez de Dios, P.-E. Forssén, K. Nordberg, and A. Ollero. Vision-based multi-UAV position estimation. *IEEE Robotics and Automation Magazine*, 13(3):53–62, 2006.
27. H. Moravec. Certainty grids for sensor fusion in mobile robots. *Sensor Devices and Systems for Robotics*, pages 243–276, 1989. Also CMU Robotics Institute 1987 Annual Research Review, 1988, pp. 33-48. Also in AI Magazine v9(2), Summer 1988, pp 61-77.
28. E. Nettleton, P.W. Gibbens, and H. Durrant-Whyte. Closed form solutions to the multiple platform simultaneous localisation and map building (SLAM) problem. In *Sensor fusion: Architectures, algorithms, and applications IV*, pages 428–437, 2000.
29. Stepán Obdrzálek and Jirí Matas. Object recognition using local affine frames on distinguished regions. In *13th BMVC*, pages 113–122, September 2002.
30. A. Ollero, J. Ferruz, F. Caballero, S. Hurtado, and L. Merino. Motion compensation and object detection for autonomous helicopter visual navigation in the comets system. In *Proceedings of the International Conference on Robotics and Automation, ICRA*, pages 19–24. IEEE, 2004.
31. A. Ollero and L. Merino. Control and perception techniques for aerial robotics. *Annual Reviews in Control*, (28):167–178, 2004. Elsevier (Francia).

32. L.-L. Ong, B. Upcroft, M. Ridley, T. Bailey, S. Sukkarieh, and H. Durrant-Whyte. Decentralised data fusion with particles. Australasian Conference on Robotics and Automation, 2005.
33. M. Rosencrantz, G. Gordon, and S. Thrun. Decentralized sensor fusion with distributed particle filters. In *Proc. Conf. Uncertainty in Artificial Intelligence*, 2003.
34. S. Russell and P. Norvig. *Artificial Intelligence: A Modern Approach.* Prentice Hall, second edition, 2003.
35. M. Saptharishi, C.S. Oliver, C.P. Diehl, K. Bhat, J. Dolan, A. Trebi-Ollennu, and P. Khosla. Distributed surveillance and reconnaissance using multiple autonomous ATVs: CyberScout. *IEEE Transactions on Robotics and Automation*, 18(5):826 – 836, Oct. 2002.
36. T. Schmitt, R. Hanek, M. Beetz, S. Buck, and B. Radig. Cooperative probabilistic state estimation for vision-based autonomous mobile robots. *IEEE Transactions on Robotics and Automation*, 18:670–684, October 2002.
37. B. Schumitsch, S. Thrun, G. Bradski, and K. Olukotun. The information-form data association filter. In *Proceedings of Conference on Neural Information Processing Systems (NIPS)*, Cambridge, MA, 2005. MIT Press.
38. R. Simmons, D. Apfelbaum, W. Burgard, M. Fox, D. an Moors, S. Thrun, and H. Younes. Coordination for multi-robot exploration and mapping. In *Proceedings of the AAAI National Conference on Artificial Intelligence*, 2000.
39. K. Singh and K. Fujimura. Map making by cooperating mobile robots. In *Proceedings of the International Conference on Robotics and Automation*, pages 254–259, 1993.
40. R.D. Smallwood and E.J. Sondik. The optimal control of partially observable markov processes over a finite horizon. *Operations Research*, 21(5):1071–1088, 1973.
41. P. Stepan, M. Kulich, and L. Preucil. Robust data fusion with occupancy grid. *IEEE Transactions on Systems, Man and Cybernetics, Part C*, 35(1):106–115, February 2005.
42. L. D. Stone, T. L. Corwin, and C. A. Barlow. *Bayesian Multiple Target Tracking.* Artech House, Inc., Norwood, MA, USA, 1999.
43. A. Stroupe, M.C. Martin, and T. Balch. Distributed sensor fusion for object position estimation by multi-robot systems. In *Proceedings of the International Conference on Robotics and Automation*, May 2001.
44. A.W. Stroupe, R. Ravichandran, and T. Balch. Value-based action selection for exploration and dynamic target observation with robot teams. In *Proceedings of the International Conference on Robotics and Automation*, volume 4, pages 4190–4197, 2004.
45. S. Sukkarieh, E. Nettleton, J.-H. Kim, M. Ridley, A. Goktogan, and H. Durrant-Whyte. The ANSER Project: Data Fusion Across Multiple Uninhabited Air Vehicles. *The International Journal of Robotics Research*, 22(7-8):505–539, 2003.
46. S. Thrun, W Burgard, and D. Fox. *Probabilistic Robotics.* The MIT Press, 2005.
47. S. Thrun, Y. Liu, D. Koller, A.Y. Ng, Z. Ghahramani, and H. Durrant-Whyte. Simultaneous Localization and Mapping with Sparse Extended Information Filters. *The International Journal of Robotics Research*, 23(7-8):693–716, 2004.
48. C. Tomasi. *Shape and motion from image streams: a factorization method.* PhD thesis, Carnegie Mellon University, 1991.
49. S. Utete and H.F. Durrant-Whyte. Reliability in decentralised data fusion networks. In *Proc. of the IEEE International Conference on Multisensor Fusion and Integration for Intelligent Systems*, pages 215–221, Oct. 1994.

50. R. Vidal, S. Sastry, J. Kim, O. Shakernia, and D. Shim. The berkeley aerial robot project (bear). In *Proceeding of the International Conference on Intelligent Robots and Systems, IROS*, pages 1–10. IEEE/RSJ, 2002.
51. R. Vidal, O. Shakernia, H. J. Kim, D.H. Shim, and S. Sastry. Probabilistic pursuit-evasion games: Theory, implementation, and experimental evaluation. *IEEE Transactions on Robotics and Automation*, 18(5):662–669, Oct. 2002.
52. T. Weigel, J.-S. Gutmann, M. Dietl, A. Kleiner, and B. Nebel. CS Freiburg: coordinating robots for successful soccer playing. *IEEE Transactions on Robotics and Automation*, 18(5):685–699, October 2002.
53. Y. Yanli, A.A. Minai, and M.M Polycarpou. Evidential map-building approaches for multi-UAV cooperative search. In *Proc. of the American Control Conference*, pages 116–121, Jun. 2005.
54. Z. Zhang. Parameters estimation techniques. a tutorial with application to conic fitting. Technical report, INRIA, France, October 1995.
55. Z. Zhang. A flexible new technique for camera calibration. *IEEE Trans. Pattern Anal. Mach. Intell.*, 22(11):1330–1334, 2000.

5

Autonomous Helicopters

Volker Remuß[1], Carsten Deeg[1], Marek Musial[1], Günter Hommel[1], Manuel Béjar[3], Federico Cuesta[2], and Aníbal Ollero[2]

[1] Technische Universität Berlin, Strasse des 17. Juni 135, 10623 Berlin (Germany) `{remuss,deeg,musial,hommel}@cs.tu-berlin.de`
[2] Robotics, Vision and Control Group, University of Seville, Camino de los Descubrimientos s/n, 41092 Seville (Spain) `{fede,aollero}@cartuja.us.es`
[3] Pablo de Olavide University, Crta. Utrera km. 1, 41013 Seville (Spain) `mbejdom@upo.es`

Summary. This Chapter is devoted to autonomous helicopters and particularly to present the MARVIN autonomous helicopter and its control system. The first section of the Chapter overviews autonomous helicopters and their control architectures and existing control methods, and then presents the MARVIN helicopter system. The second section is devoted to the helicopter model. The third section deals with the helicopter control techniques. Finally, some conclusions are presented.

5.1 Autonomous Helicopters Overview

5.1.1 General Issues on Autonomous Helicopters and Control Architectures

Autonomous Helicopters

Autonomous helicopters present several advantages for aerial robotics. In fact, the manoeuvrability of helicopters is very convenient for many robotic applications. Furthermore, the ability to maintain the aerial vehicle in hovering is also important in many robotic tasks.

However, helicopters are more difficult to control and require the application of reliable control laws. Furthermore, helicopters require experienced safety pilots for their development and application.

Several Universities in the USA have developed autonomous helicopters. Thus, the Robotics Institute at Carnegie Mellon University (CMU) conducted since the early nineties an autonomous helicopter project. They have developed different prototypes from small electrical radio controlled vehicles to autonomous helicopters using the Yamaha R50 platform. The autonomous CMU helicopter won the AUVSI aerial robotic competition in 1997.

The University of Southern California (USC) conducted, since 1991, an autonomous helicopter project developing several prototypes, such as the AVATAR (Autonomous Vehicle Aerial Tracking and Retrieval/ Reconnaissance) prototypes

A. Ollero and I. Maza (Eds.): Mult. Hetero. Unmanned Aerial Vehi., STAR 37, pp. 111–146, 2007.
springerlink.com

presented in 1994 and 1997. The AVATAR helicopter won the AUVSI Aerial Robotics competition in 1994.

The University of Berkeley also developed autonomous helicopters in the Berkeley AeRobot project, BEAR, in which the autonomous aerial robot is a testbed for an integrated approach to intelligent systems.

The Georgia Institute of Technology (GIT) has the Unmanned Aerial Vehicle Research facility and developed several platforms and aerial autonomous systems during the last decade. GIT also won the AUVSI aerial robotics competition.

In Europe the University of Linköping conducted the WITAS project which was a long term basic research project involving cooperation with other Universities and private companies [10]. The Yamaha Rmax helicopter was used in the WITAS project. Moreover, several Universities such as the Technical University of Berlin, that won the 2000 Aerial Robotics Competition, ETH Zurich, Universidad Politécnica de Madrid and Universidad de Sevilla are using the adaptation of conventional radio controlled helicopters with different autonomous capabilities.

In the next section of this Chapter the MARVIN autonomous helicopter, from the Technical University of Berlin (TUB), will be presented.

Control Architectures

The position and orientation of an helicopter (six degrees of freedom) are usually controlled by means of 5 control inputs: the main rotor thrust (power to the rotor), the collective input (main rotor blade angle) which has a direct effect on the helicopter height (altitude control), the tail rotor which affects the heading of the helicopter (yaw motion) and compensates the anti-torque generated by the main rotor, the longitudinal cyclic which modifies the helicopter pitch angle and the longitudinal translation, and the lateral cyclic, which affects the helicopter roll angle and the lateral translation (lateral cyclic). The helicopter is a multivariable non-linear underactuated system with strong coupling in some control loops.

The University of Southern California (USC) developed a behavior-based architecture for the control of the AVATAR helicopter [13]. The low-level behaviors correspond to the generation of the four input commands of the helicopter (collective throttle, tail rotor, longitudinal and lateral cyclic). The second level implements short-term goal behaviors: transition to altitude and lateral velocity. The highest-level behavior, navigation control, is responsible for long-term goals such as moving to a particular position and heading.

Intelligent control architectures for unmanned air vehicles (helicopters) have been also researched at Berkeley. The hierarchical architecture segments the control tasks into different layers of abstraction in which planning, interaction with the environment and control activities are involved. The hierarchical flight management system [25] has a stabilization/tracking layer, a trajectory generation layer, responsible for generating a desired trajectory or a sequence of flight modes, and a switching layer which switches between several strategy planners. Both continuous and discrete event systems are considered. In order to model these control systems, hybrid system theory has been proposed (see for example [29]).

GTI also developed autonomous helicopter control systems and research in flight controls, avionics and software systems.

Learning and Pilot Knowledge-Based Control Methods

As far as the research in control methods is concerned, different approaches can be used. Thus, fuzzy logic has been applied to control the Yamaha's helicopter at the Tokyo Institute of Technology, which demonstrated autonomous capabilities and also person-machine interfaces including voice command [35].

Fuzzy logic with rules generated by the observation of a human pilot and consultation with helicopter experts is the approach used in [5].

In [41] the behaviors of the control architecture proposed in the USC architecture are implemented as PD control loops with gains tuned by trial and error. In [40], the "teaching by showing" approach is presented. In this work the controller is generated by using training data gathered while a human teacher controls a system until the synthesized controller can also control the system to meet predefined performance criteria.

In [4] learning is based on the direct mapping of sensor inputs to actuator control via an artificial neural network. Then, the neural network controller was used for the helicopter hovering.

The analysis of the pilot's execution of aggressive maneuvers from flight test data is the base of the method presented in [16] to develop a full-non-linear dynamic model of a helicopter. This model will be used in the design of new control systems for autonomous helicopters.

Model Based Control Methods

On the other hand, several methods have been applied for model based control of UAVs. Modelling the UAV dynamics is a main issue. The full model of a helicopter involving the flexibility of the rotors and fuselage and the dynamics of the actuators and the combustion engine is very complex. Then, in most cases, the helicopter is considered as a rigid body with inputs forces and torques applied to the centre of mass and outputs the position and linear velocities of the centre of mass, as well as the rotation angles and angular velocities. Furthermore, the relations between the control inputs of the helicopter and the above mentioned forces and torques should be considered in the model. In general, these relations involve the aerodynamics of the fuselage and the effect of stabilizers. However, at low speeds these effects can be ignored [31].

In [26] a mathematical model and experimental identification of a model helicopter is presented. The model of the interactions between the stabilizer flybar and the main rotor blade is also included showing its effects in the stability of the model helicopter. The identification of the parameters is performed on a SISO basis, using a specially-built stands to restrict the motion of the helicopter to one degree of freedom. The identification from input-output data, collected when a human pilot is controlling the vehicle, is difficult because it is not possible to study the individual effect of each control input (the pilot has to apply more than one input to maintain the stability).

In [38] a parameterized model of the Yamaha R-50 autonomous helicopter is identified using frequency domain methods. The stabilizer bar is also taken into account. The model was validated with special flight experiments using doublet-like control inputs in hover and forward flight, showing its ability to predict the time domain response of the helicopter to control inputs.

In Sect. 5.2 of this Chapter the mathematical model of the MARVIN autonomous helicopter will be presented.

It should be noted that in hovering, the nonlinear system can be linearized and then multivariable linear control techniques such as LQR and H_∞ can be applied. In [25] multiloop linear PID techniques also obtained good results when applied to the Yamaha R-50. However, if large perturbations should be compensated, or significant tracking abilities are required, this strategy could be not enough. In this case further improvements can be obtained by adding nonlinear control terms that compensate significant deviations with respect to the hovering conditions.

In [23] a fuzzy gain-scheduling approach, based on the linearization of the original nonlinear helicopter model, is proposed and tested in simulation.

Reference [21] combines the helicopter attitude inner control loop and the outer trajectory control loop and apply adaptive techniques to cancel model errors by preventing unwanted adaptation to actuator limits and dynamics in the inner loop.

In [48] linear robust multivariable control, fuzzy logic control and nonlinear tracking control are compared in the simulation of two scenarios: vertical climb and simultaneous longitudinal and lateral motion. It is noted that nonlinear control techniques by applying feedback linearization are more general and cover wider ranges of flight envelopes but requires accurate knowledge about the system and are sensitive to model disparities, such as changes in the payload, or to the aerodynamic thrust-torque model.

In general no guarantee of robustness against model uncertainties or disturbances and no adaptive capabilities are provided by many feedback linearization techniques. However, in some cases, nonlinear controller robustness properties are increased using sliding mode and Lyapunov based control [37]. Typically, these techniques trade the controller performance against uncertainty, but require a priori estimates of parameter bounds, which may be difficult to obtain.

However, research efforts to design new robust non linear control laws are pursued. Then in [20] the vertical motion of a nonlinear model of a helicopter to a reference signal, while stabilizing the lateral and longitudinal position and maintaining a constant attitude, is studied. The problem is motivated by the synchronization of the vertical motion of the helicopter with a sea carrier subject to wave-induced oscillations, and then the reference signals are sum of sinusoidal signals (assumed not to be available to the controller). A nonlinear adaptive output regulation and robust stabilization of systems in feed-forward form by means of saturated control is applied in simulation. The simulation results show robustness against uncertainties on the model and on the exogenous reference signal. The method also requires the a priori computation of robustness bounds.

In [25] the application of nonlinear model predictive control is proposed. At each sample time, the controller computes a finite control sequence which minimizes a quadratic index. This index includes the errors of the outputs (helicopter coordinates and heading) with respect to desired trajectories, additional state variables which should be bounded, and the control actions (longitudinal and lateral cyclic pitch, collective, and tail rotor). A gradient descent technique is used to compute the optimal values of the control variable. The method improves the tracking performance at the expenses of heavy computing load.

In [14] the control of underactuated systems including helicopters and Planar VTOL (PVTOL) is studied. Several control techniques are presented including backstepping, energy based controllers and Lyapunov-based controllers.

At CMU a high-order linear model of the R-50 Yamaha helicopter is used for control. This model was extracted by using the MOSCA (MOdelling for Flight Simulation and Control Analysis) with a non-linear simulation model of the helicopter. The controller consists of one multivariable (MIMO) inner loop for stabilization and four separate (SISO) guidance loops for velocity and position control. Several manoeuvre tests have been conducted with the helicopter (square, forward turn, backward turn and nose-out circle). The controller is designed for hovering but its robustness leads the helicopter to perform the manoeuvres efficiently even if the trajectories are not optimal [37].

The control architecture of the MARVIN helicopter is also described in Sect. 5.3 of this Chapter. Different control techniques to the MARVIN autonomous helicopter are applied.

5.1.2 The MARVIN Autonomous Helicopter System

MARVIN (multi-purpose aerial robot vehicle with intelligent navigation) Mark II (M2) is an autonomous unmanned aerial vehicle (UAV) that was built in 2005 (see Fig. 5.1).

Its predecessor [42] was developed from 1997 to successfully win the International Aerial Robotic Competition (IARC) [15] in 2000. From 2002 to 2005 it was used in cooperation with other UAVs in the European research project COMETS [7].

The M2 Airframe

The M2 airframe is a commercially available helicopter made in Germany by Aero-Tec [2]. Its main feature is the sealed, full metal, single-stage main rotor gear which is mounted directly on the engine, building a self-supporting unit (see Fig. 5.2).

Its additional features make it a good choice for an autonomous helicopter airframe. The setup acquired by TUB is equipped with a magnetic sensor on the main rotor axis to be able to measure the rotor's rotation per minute (r p m), a high-stand landing gear with mounting capabilities and an electric starter.

Fig. 5.1. MARVIN mark II in autonomous flight

Fig. 5.2. Airframe's mechanics

The remote control (RC) equipment and gyro uses standard parts produced by Graupner [24]. The engine is a Zenoah G230RC two-stroke 23 cm^3 gasoline-oil-mix engine with a maximum power output of 1.84 kW at 12500 r p m [6].

The helicopter has a two-bladed main rotor with a diameter of 1.84 m and a two-bladed tail rotor with a diameter of 0.34 m. The tail rotor is coupled to the main rotor via a fixed-ratio gear. The main rotor's rotary speed is actively fixed during flights at about 1350 r p m. Changes in lift, which are needed to move the helicopter, are produced by changing the main rotor blades' pitch. This is done symmetrically (collective pitch) to change the lift and asymmetrically (cyclic pitch) to accelerate in a direction. To set the pitch relative to the helicopter a so-called swashplate (SP) is used. It is moved by three servos. The tail rotor is used to compensate for the main rotor and engine torque and changes the heading of the helicopter. Its (collective) pitch is controlled by a single servo. A fifth servo moves the throttle. The SP and the throttle's servo can be seen in Fig. 5.2. The three SP servos are inside the dark front box of the helicopter, but their linkage is visible.

Remote-Controlling a Model Helicopter

The servos are controlled by pulse width modulated (PWM) signals where the PW is proportional to the servo anchor's position. In radio controlled flight the servos are driven by the RC receiver as received from the RC. The RC pilot does not control directly the servos but uses a higher-level approach. The pilot has two 2-way sticks, one is controlling the cyclic pitch, which has 2 degrees of freedom (DOF) and the other one is controlling the collective and the tail rotor pitch with 1 DOF each. The throttle is usually fixed to a certain base level. The RC derives the individual servo signals from the pilot's commands. The RC also compensates for some couplings, i.e. it raises throttle when the collective pitch is raised. In case of the M2 airframe the level of abstraction is very low because there is a dedicated collective pitch servo that moves the other two (cyclic pitch) servos.

Mostly, an RC pilot uses a 1 DOF gyroscope to control the tail rotor pitch to prevent unwanted rotation around the main rotor axis. All of these automatisms have to be adjusted for each particular set of helicopter model, servos, rotor blades, fuel, weather and so on. The RC transmits unidirectional, so there is nothing but visual feedback from the flying helicopter.

From a Remote-Piloted Helicopter to an Autonomous UAV

To be able to fly a model helicopter autonomously one has to control the helicopter automatically. The automatic control can be engaged on any point between in-front-of-the-remote-control to just-before-the-servos. Due to automatisms mentioned above, the decision was taken to go to the lowest level and directly interface the servos on the helicopter and to spare out any RC automatisms. A microcontroller (MC) was used for this purpose. The decision to switch between autonomous- and remote-controlled modes is taken by the MC's program but can be overridden by a lever on the RC for security reasons. Switching back to manual flight is always the last resort in case of system problems.

Since two different batteries for the microcontroller and the RC receiver and servos are used, the hardware falls back to remote-controlled flight if the MC is running out of battery. In addition the MC is able to monitor the signals that are sent to the servos during a remote-piloted flight thus they can be logged. In earlier versions of MARVIN all the servo and receiver interfacing was performed using dedicated custom-built hardware and the MC directly. In M2, this is no longer necessary since a receiver system is available that comes with this functionality built-in (see below). M2 can also start its engine autonomously using the electric on-board starter.

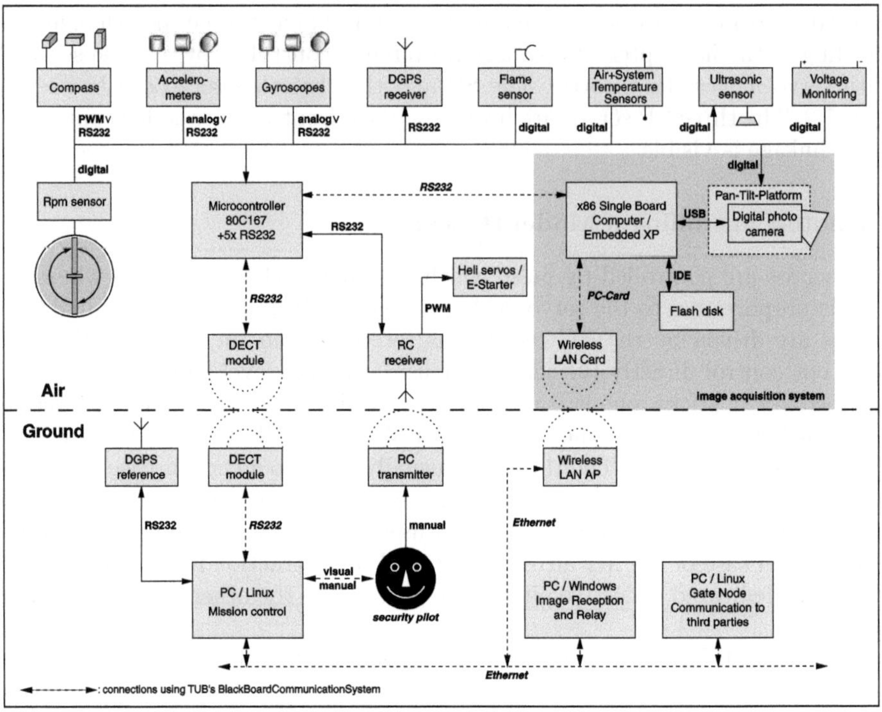

Fig. 5.3. Complete autonomous M2 system incl. ground base and optional image acquisition system.

The Infineon C167 MC is the only CPU on board the helicopter used for autonomy. Therefore, all the needed sensors are connected to it and it executes all programs for control purposes, sensor data acquisition and calculations. To fly the helicopter autonomously the control program in the microcontroller needs information about the current position and the attitude of the helicopter. Additionally, the current main rotor rotations per minute (r p m) are needed to control the engine's throttle and a sensor for the distance to the ground for landing purposes.

The system uses a Differential Global Positioning System (DGPS) to get accurate and frequent position information and a combination of three rotational speed, three acceleration and three magnetic field sensors for movement detection. A fusion algorithm is used to calculate the helicopter's attitude from raw sensor data [43]. The main rotor r p m is measured magnetically at the rotor axle and the distance to the ground by an ultrasonic sensor. Each of these sensors is directly connected to the MC. To add the D to GPS a stream of correction data from a base station is needed. In case of M2 it is embedded in the normal data transfer between the base station and the helicopter and then transferred through the MC to the on-board GPS using TUB's real-time communication system (CS) [46, 47].

Figure 5.3 depicts the full M2 embedded architecture including a view of the ground station which consists of several networked computers using the same CS in all nodes. The architecture includes the image acquisition system as well as all components that are needed for autonomous flight.

Hardware Components Realisation

The decision was taken to go for a flexible and comfortable approach (for research, development and maintenance) and to attach a box beneath the helicopter that can contain a variable number of computer systems, sensors, power supplies, and batteries. Another benefit of this design is the quick replaceability of the airframe for repairing issues. The box is completely cut from carbon fiber sheets besides the aluminium back that is used for most connectors and as power supply's heat sink. Figure 5.4 depicts M2's integrated electronic box including the image acquisition system. Most components of the M2 system are visible.

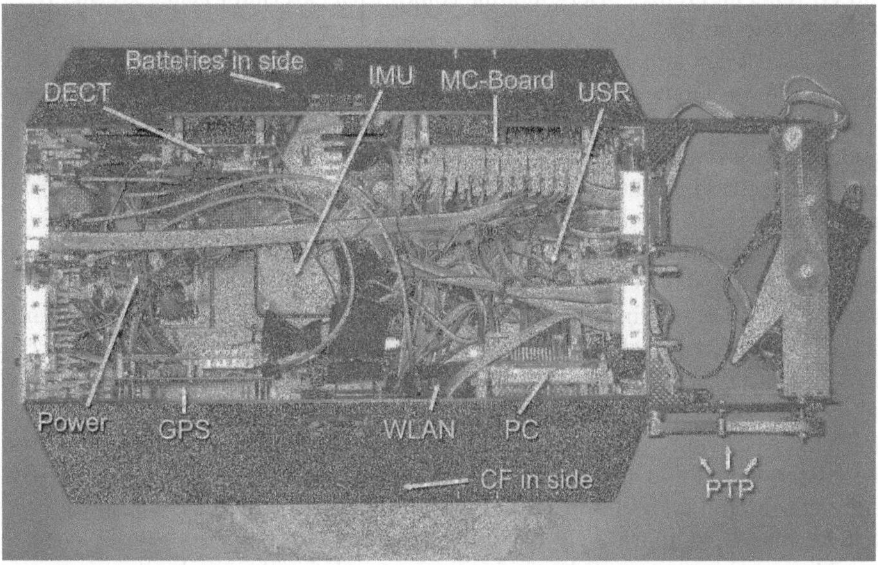

Fig. 5.4. M2's integrated electronics

The M2 uses a DECT (Digital Cordless European/Enhanced Telephone) OEM module by Höft & Wessel [3] as a transparent serial link for communication with the base station. To ease the integration an end-user product was used (HW8615). The good thing about this DECT modem is that it comes including an external antenna connector and an own power supply that fulfils the specification of the OEM module. Compared to the formerly used Siemens M101 Data, these modules were easier to integrate and use, but their actual performance was less stable.

The box in Fig. 5.4 shows an inertial measurement unit (IMU) GX1 from Microstrain [39]. This IMU integrates all needed sensors (rotation, acceleration and magnetic in 3D) and interfaces the MC via RS-232. While the integration of all these sensors in a single device seems a very nice idea at a first glance it turned out that for M2 it was not possible to find a mounting place that was suitable for all the sensors at the same time. The accelerometers and gyroscopes are best placed decoupled below the helicopter's centre of gravity while the magnetic sensors are heavily disturbed by the two-stroke engine's induction coil and are best placed as far away as possible. In M2's case "far away" is the helicopter's tail section. So it was necessary to add a compass to the Microstrain when it was mounted as in the picture above or to use the Microstrain as compass only (mounted on the tail) and use another IMU in the box.

Heart of the system is the custom made MC-motherboard that contains an integrated Infineon C167 MC module by Frenzel+Berg [12], temperature and voltage monitoring and a lot of lockable connectors. The C167 is a 16-Bit microcontroller of good connectivity including Capture/Compare-Units, AD-Converters and GPIO. The Frenzel+Berg module comes with external Flash, external RAM and 4 additional RS-232 serial ports.

The used GPS is a high accuracy OEM-4G2L by Novatel [18]. This GPS is used in differential mode accompanied by a second, older Novatel receiver (OEM-3) on ground. A second existing M2 helicopter is also using the older OEM-3 model. Both systems show a similar accuracy of 1 cm while the OEM-4 has a maximal position update rate of 20 Hz and the OEM-3 of 5 Hz. The M2 system uses 5 Hz or 10 Hz, respectively.

Directly integrated into the base plate of the box facing downwards is an ultrasonic rangefinder (USR) produced by SmartSens [19]. It is mainly used to detect the ground during autonomous landing. Four high-bright red light emitting diodes (LEDs) in the base plate are used to give feedback about the current flight mode (i.e. helicopter turning, hovering, start accelerating, or landing) when the helicopter is autonomous.

The power supply unit at the aluminium sheet takes a variable voltage from the batteries and converts it to 5 V and 3.3 V using mostly switching DC-DC-Converter modules PL10S produced by Lambda [32]. These modules produce a well regulated, comparatively low-noise output suitable for all digital systems. In addition, there are very low-noise linear regulators for analogue systems especially sensors and AD-converters. The complete M2 system is powered using lithium-polymer rechargeables batteries from Kokam [27]. A 11.1 V, 3 Ah, 235 g battery pack powers the system including the image acquisition system for at least 90 minutes. The M2 can carry up to 3 batteries which are automatically balanced during discharge.

Not inside the box is the DDS-10 RC-receiver produced by ACT [1]. Since the receiver is used by the helicopter for manual RC it is installed in the helicopter's nose and powered by a dedicated battery. This receiver has a unique feature called diversity synchro-link system (DSL-S) which enables the C167 to control the servos via RS-232 or read the values as sent by the RC. A master-slave setup still enables a security override switch at the RC. The receiver's design is quite robust, it falls automatically back to RC when the microcontroller stops sending datagrams but it also keeps on using the microcontroller's data in case the RC is failing or out of reach. Additionally, the receiver monitors its current reception strength and battery voltage.

Also in the box is the image acquisition system's embedded PC. This PC is a PC-104+ compliant, embedded National Semiconductor (now AMD) Geode system-on-chip SB-i686 by Compulab [34]. It is operated with embedded Windows XP to drive the digital still camera that is connected via USB and boots from a compact flash card (CF) attached via IDE. Its pc-card slot carries a wireless LAN card for image transfer to the base station, but the wireless LAN can also transmit all data transferred from the MC to the base station in case of a DECT failure. The embedded PC is not necessary for autonomous flight. Together with the WLAN, CF, the pan-tilt-platform (PTP) and the camera it is just payload.

Another mission-specific sensor is the flame sensor hidden in a small external white box at the front most part of the helicopter. This fire or more exact a flame sensor is a Hamamatsu UV-Tron R2868 [44]. It is very sensitive in detecting flames but is absolutely insensitive to other sources of heat and light, e.g. the sun. The sensor is detecting the intensity. In the M2 system the sensor is masked to a narrow opening angle in the horizontal plane and it is looking to the front, in flying direction. In this configuration it can be used to avoid flying into fires and to triangulate them.

Base Station

The M2 base station consists of several components but the salient program is the Mission Control, mainly because it has a GUI (see Fig. 5.5) and represents the base station's user-view.

This program can be used to plan and survey a mission flight including definition of waypoints to fly to and giving the command for start and abortion if necessary. It can display some basic digital map data, the helicopter's position and heading as well as the flown path. Some status information and flight parameters are shown as well. Behind the scenes the MC sends atomic commands to the helicopter systems that are then executed independently. Terminating the MC does not end the autonomous flight.

MARVIN's Flight

Flying a path for MARVIN is always flying from one three-dimensional waypoint $(i-1)$ to another (i). MARVIN has two different flight modes. One is *straight line* and the other *curved*.

When flying *straight line* mode, the system takes the shortest path from waypoint $i-1$ to i usually nose ahead. If i is reached, it stops, turns toward $i+1$ and accelerates. MARVIN is flying the way with a given (maximal) cruise speed.

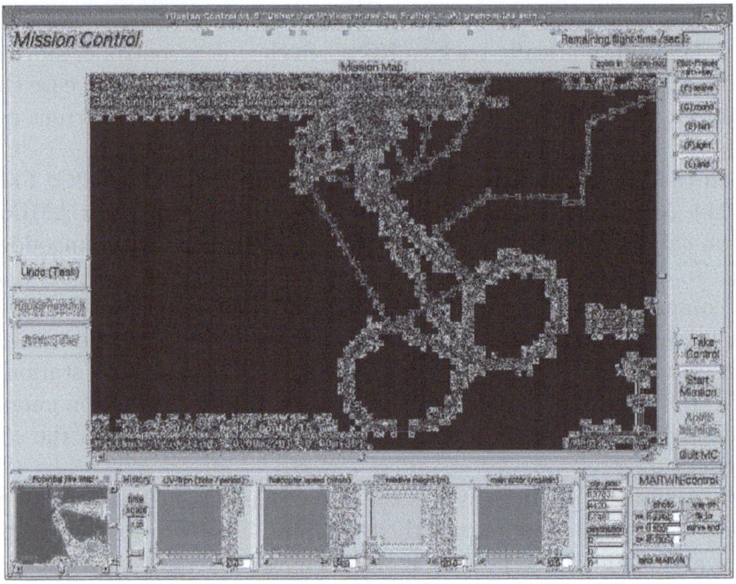

Fig. 5.5. Mission Control Program (view after a real flight)

When flying *curved* mode, the helicopter keeps the cruise speed until it reaches the last way-point and does not stop at every intermediate waypoint. To achieve that, the control switches from waypoint i to $i+1$ as soon as the helicopter is so close to i that it would need to slow down to stop at point i. The result is a smooth transition between waypoints. Figure 5.6 depicts a simulated flight that shows both modes in comparison.

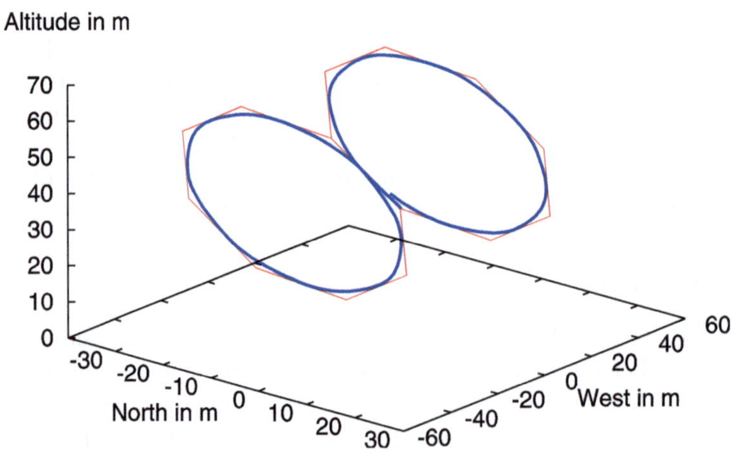

Fig. 5.6. Simulated flight

It is possible but rarely used to fix the direction of MARVIN's nose while flying. The mission flight speed is typically 2 to 5 m/s, but successful flights have been performed with a velocity of 14 m/s.

The flight range of the helicopter in its current state is about 300 m due to two issues. Firstly, as long as a human pilot should be able to recover the helicopter in case of an emergency it has to be somewhere in the middle of the area because the visual range is about 150 m. Secondly, DECT and WLAN are officially limited to 300 m but this is solely a monetary issue which could be solved easily.

The maximum flying height is an even more difficult question. A maximal ascending speed of 2 m/s can be reached, but flying beyond a height of 50 m is risky since this is the maximum height that a human pilot is able to fly. Disregarding the safety factor, the reachable height is 300 m as well due to the radio components.

Safety is an issue. It may seem a pretty small helicopter but if one looks at the technical data and think about it, one will see that there are rotating blades with 1350 r p m, having a diameter of 1.84 m and weight 0.25 kg each. Therefore, the tip of the blade has a speed of more than 450 km/h and each blade is pulling with a force of about 2800 N outwards.

5.2 Helicopter Model

Obtaining a mathematical model is important for designing the control system as well as for simulation of a helicopter. The mathematical model presented below does not describe all the mechanical parts of a helicopter as for example in [26, 38] (especially the mechanics of the flybar). Thus, the resulting model as published in [9] can easily be applied to other helicopters with different mechanics. On the other hand, this is a simplification but compared to real flight tests the results are good enough for many applications. A discussion of the complete detailed model including elaborated treatment of the rotor's aerodynamics as derived for MARVIN can be found in [8].

5.2.1 Coordinate Systems

The model uses two different coordinate systems:

- The *base coordinate system* (BCS) serves as the fixed Newtonian reference frame for navigation way-points. It is defined by the directions of geographic North (x_{BCS}) and vertical up (z_{BCS}), see Fig. 5.7. The origin of the BCS can be chosen by the user, one obvious choice being the position of the GPS reference antenna.
- The *helicopter coordinate system* (HCS) is fixed to the helicopter and is defined as x_{HCS} for "forward" and z_{HCS} for "rotor axis up", the origin in the rotor axis and close to the vehicle's center of mass. See Fig. 5.8.

Unless otherwise noted, all coordinates throughout this section will refer to the HCS.

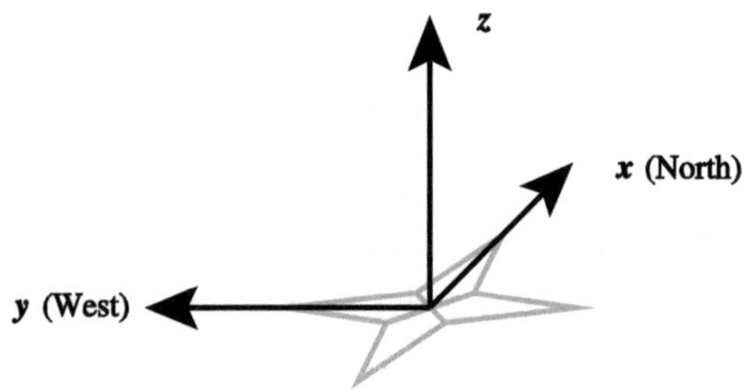

Fig. 5.7. Base Coordinate System (BCS)

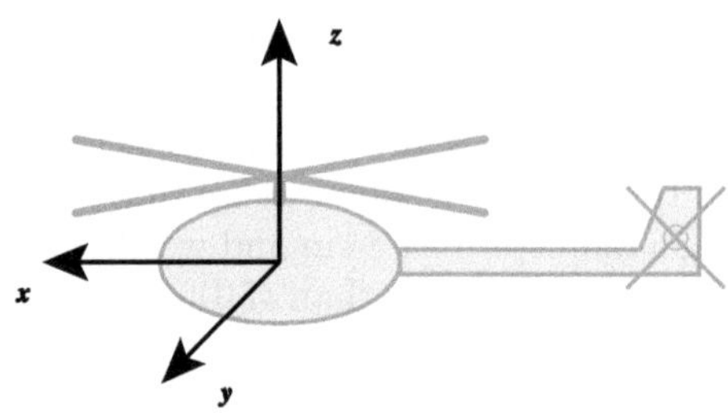

Fig. 5.8. Helicopter Coordinate System (HCS)

5.2.2 Helicopter State

The state $\mathbf{x}$ of the helicopter describes the location, velocity and all the other values (see Table 5.1) that change over time and which are needed for calculating future states. The update of the state can be calculated using the standard system equation from a function $\mathbf{f}$ of the state and the input $\mathbf{u}$ to the system:

$$\dot{\mathbf{x}} = \mathbf{f}(\mathbf{x}, \mathbf{u}) \tag{5.1}$$

The derivation of $\mathbf{f}$ is described in this section.

Table 5.1. State of a helicopter

$\mathbf{s}$, $\mathbf{v} = \dot{\mathbf{s}}$	position, velocity of the center of gravity
φ, $\omega = \dot{\varphi}$	orientation, angular velocity
ω_M	angular velocity of main rotor
p_c, p_x, p_y	main rotor collective and cyclic pitch
p_t	tail rotor (collective) pitch
th	throttle for engine

5.2.3 Physics of Motion

For the simulation of movement, kinematic equations are needed that describe the alteration of position and velocity (called the state of the helicopter) with respect to the current state and applied forces. During simulation, these equations are numerically integrated to compute the changing helicopter state during simulation time.

Translational movement is described by the simple and well known equation of Newton (applied force $\mathbf{F}$, mass of the helicopter m).

$$^{\mathrm{BCS}}\mathbf{F} = m^{\mathrm{BCS}}\ddot{\mathbf{s}} \tag{5.2}$$

Rotational movements are described by the conservation of angular momentum $\mathbf{L}$. With the resulting torque $\mathbf{M}$ of all simulated effects (see below), the corresponding equation can be written as

$$^{\mathrm{BCS}}\mathbf{M} = {}^{\mathrm{BCS}}\dot{\mathbf{L}} \tag{5.3}$$

Since this equation is only valid in an inertial frame such as BCS, it has to be transformed to be used in HCS. According to [36] the result is the so called *Euler equation*:

$$^{\mathrm{HCS}}\mathbf{M} = ^{\mathrm{HCS}}\dot{\mathbf{L}} + {}^{\mathrm{HCS}}\boldsymbol{\omega} \times {}^{\mathrm{HCS}}\mathbf{L} \tag{5.4}$$

In the following equations the index HCS is omitted.

The momentum can be divided into three main parts: $\mathbf{L}_H$ of the helicopter, $\mathbf{L}_M$ of the main rotor and $\mathbf{L}_T$ of the tail rotor.

$$\mathbf{L} = \mathbf{L}_H + \mathbf{L}_M + \mathbf{L}_T \tag{5.5}$$

For calculation of the angular momenta they have to be divided into moment of inertia and angular velocity. For the main rotor this can easily be written as $\mathbf{L}_M = J_M \boldsymbol{\omega}_M$ with the moment of inertia J_M. The tail rotor is treated identically. Even the rotating parts of the engine can be modelled this way (index E).

The helicopter body can rotate around every axis, so that its moment of inertia has to be interpreted as a tensor J_H. This tensor incorporates the distribution of mass of the fuselage and additionally the distribution of mass of the rotors. The latter is assumed to be uniformly distributed over the whole rotor disk. This leads to the complete description of angular momentum:

$$\mathbf{L} = \mathsf{J}_H \cdot \boldsymbol{\omega} + J_M \boldsymbol{\omega}_M + J_T \boldsymbol{\omega}_T + J_E \boldsymbol{\omega}_E \tag{5.6}$$

For simplification of future equations the direction of rotation is assigned to the moments of inertia (e.g. $\mathbf{J}_M := J_M \frac{\boldsymbol{\omega}_M}{|\boldsymbol{\omega}_M|}$). Since all rotating parts are coupled with fixed ratios, it is possible to define $\omega_M > 0$ as reference angular velocity. ω_T and ω_E can be calculated using fixed ratios (e.g. $n_T := \frac{|\boldsymbol{\omega}_T|}{|\boldsymbol{\omega}_M|}$):

$$\mathbf{L} = \mathsf{J}_H \cdot \boldsymbol{\omega} + (\mathbf{J}_M + n_T \mathbf{J}_T + n_E \mathbf{J}_E)\omega_M \tag{5.7}$$

The final equation of rotational movement of the helicopter can be composed from equations (5.4) and (5.7).

$$\mathbf{M} = \mathsf{J}_H \cdot \dot{\boldsymbol{\omega}} + (\mathbf{J}_M + n_T \mathbf{J}_T + n_E \mathbf{J}_E)\dot{\omega}_M + + \boldsymbol{\omega} \times (\mathsf{J}_H \cdot \boldsymbol{\omega} + (\mathbf{J}_M + n_T \mathbf{J}_T + n_E \mathbf{J}_E)\omega_M) \tag{5.8}$$

Equation (5.8) was derived without simplifications and can be used for simulation. But if this equation has to be simplified for some reason (e.g. deriving a controller law), this is possible, since ω_M is much bigger than ω and almost constant. Additionally $n_T\mathbf{J}_T$ and $n_E\mathbf{J}_E$ are small compared to $\mathbf{J}_M$. This leads to

$$\mathbf{M} \approx \begin{cases} J_H \cdot \dot{\boldsymbol{\omega}} & \text{if} \quad \boldsymbol{\omega}\Box\text{HCS}_z \\ \boldsymbol{\omega} \times \mathbf{J}_M\omega_M & \text{if} \quad \boldsymbol{\omega}\perp\text{HCS}_z. \end{cases} \tag{5.9}$$

The result describes precession movement of a spinning gyro for pitching and rolling movement around HCS-x and HCS-y axes. Due to the cross product this movement is shifted by 90 degrees with respect to the applied torque $\mathbf{M}$. Additionally $\mathbf{M}$ is proportional to the angular velocity ω and not to the acceleration as stated in [26]. In [17, 26] the effects of the spinning rotor are completely neglected. Reference [28] shows that the contribution of these effects makes up more than 99% and therefore must not be neglected.

In order to solve equation (5.8) or (5.9), ω_M is needed. The rotational speed depends on the torque of the engine M_E and drag of the rotors, which can be described as torque of the rotor in rotation direction (e.g. $\mathbf{M}_M \cdot \frac{\mathbf{J}_M}{|\mathbf{J}_M|}$). Friction inside the engine is already considered in the torque plot of the engine, which can be used to determine M_E. Since the engine drives the rotors through their axes, only the movement parallel to these axes has to be considered:

$$\begin{aligned} M_E - \mathbf{M}_M \cdot \frac{\mathbf{J}_M}{|\mathbf{J}_M|} - \mathbf{M}_T \cdot \frac{\mathbf{J}_T}{|\mathbf{J}_T|} &= \sum_{i\in\{M,T,E\}} J_i^{\text{BCS}}\dot{\boldsymbol{\omega}}_i \cdot \frac{^{\text{BCS}}\boldsymbol{\omega}_i}{|^{\text{BCS}}\boldsymbol{\omega}_i|} = \\ &= \sum_{i\in\{M,T,E\}} J_i(^{\text{HCS}}\dot{\boldsymbol{\omega}}_i + {}^{\text{HCS}}\dot{\boldsymbol{\omega}}) \cdot \frac{^{\text{HCS}}\boldsymbol{\omega}_i}{|^{\text{HCS}}\boldsymbol{\omega}_i|} = \\ &= \sum_{i\in\{M,T,E\}} (\mathbf{J}_i \cdot {}^{\text{HCS}}\dot{\boldsymbol{\omega}} + |\mathbf{J}_i|\dot{\omega}_i) = \\ &= (\mathbf{J}_M + \mathbf{J}_T + \mathbf{J}_E) \cdot \dot{\boldsymbol{\omega}} + (|\mathbf{J}_M| + n_T|\mathbf{J}_T| + n_E|\mathbf{J}_E|)\dot{\omega}_M \end{aligned} \tag{5.10}$$

5.2.4 Forces and Torques

Equations (5.2), (5.8) and (5.10) have to be solved for $\ddot{\mathbf{s}}$, $\dot{\boldsymbol{\omega}}$ and $\dot{\omega_M}$. They depend on the sums of all applied forces $\mathbf{F}$ and torques $\mathbf{M}$, described below.

$\mathbf{F}$ is the sum of all forces created by the two rotors ($\mathbf{F}_M$, $\mathbf{F}_T$) and air resistance of the fuselage ($\mathbf{F}_F$). Finally, the helicopter's gravity $^{\text{BCS}}\mathbf{F}_g = (0, 0, mg)^T$ (with acceleration of gravity g) has to be taken into account.

$$\mathbf{F} = \mathbf{F}_M + \mathbf{F}_T + \mathbf{F}_F + \mathbf{F}_g \tag{5.11}$$

The calculation of $\mathbf{F}_M$ and $\mathbf{F}_T$ is described below. $\mathbf{F}_F$ has to be calculated from the air speed relative to the helicopter $\mathbf{v}_\omega$, which consists of wind $\mathbf{w}$ and the movement of the helicopter itself, which gives $\mathbf{v}_\omega = \mathbf{w} - \mathbf{v}$.

As it can be found in any book about aerodynamics (e.g. [33]), air resistance can be calculated using density of air ρ, the resistance coefficient C_ω and related cross-sectional area together with relative air speed. This is done for each direction (the absolute value is used to obtain the force in the correct direction):

$$\mathbf{F}_F = \frac{1}{2}\rho \begin{pmatrix} C_{wx}\ A_{Hx}\ v_{wx}\ |v_{wx}| \\ C_{wy}\ A_{Hy}\ v_{wy}\ |v_{wy}| \\ C_{wz}\ A_{Hz}\ v_{wz}\ |v_{wz}| \end{pmatrix} \tag{5.12}$$

The overall torque $\mathbf{M}$ contains all torques produced by both rotors ($\mathbf{M}_M$, $\mathbf{M}_T$). Additionally the forces $\mathbf{F}_M$, $\mathbf{F}_T$ of both rotors produce a torque, if they are displaced from the center of gravity ($\mathbf{P}_M$, $\mathbf{P}_T$). Finally the air resistance force $\mathbf{F}_F$ can also produce a torque $\mathbf{M}_F$, if this force is not applied to the center of gravity.

$$\mathbf{M} = \mathbf{M}_M + \mathbf{M}_T + \mathbf{P}_M \times \mathbf{F}_M + \mathbf{P}_T \times \mathbf{F}_T + \mathbf{M}_F \tag{5.13}$$

With $\mathbf{P}_{Fi}$ as point of application of F_{Fi}, $\mathbf{M}_F$ can be calculated:

$$\mathbf{M}_F = \mathbf{P}_{Fx} \times \begin{pmatrix} F_{Fx} \\ 0 \\ 0 \end{pmatrix} + \mathbf{P}_{Fy} \times \begin{pmatrix} 0 \\ F_{Fy} \\ 0 \end{pmatrix} + \mathbf{P}_{Fz} \times \begin{pmatrix} 0 \\ 0 \\ F_{Fz} \end{pmatrix} \tag{5.14}$$

5.2.5 Aerodynamics of a Rotor

This section describes the aerodynamic equations for the main rotor, but they are applied to the tail rotor as well. This model relies on Blade Element Theory only, which is described in [33, 22, 11]. No induced wind velocities are taken into account.

Forces of a single rotating blade are not constant during one cycle of the rotor. Due to wind and cyclic pitch, lift and drag change during one cycle. For control purposes, these forces and the resulting torques can be combined to mean values. MARVIN has a main rotor speed of approximately 1300 r p m and two blades, which gives an effective frequency of 40 Hz. That means it is not reasonable to increase the controlling frequency of the used controller beyond this frequency using just mean values of the forces.

For determination of mean values of forces, all components of each rotor element have to be integrated according to Fig. 5.9. The blades are assumed to be non-twisted and of constant chord from radius R_1 to R_2.

For calculation of the force $\mathrm{d}F$ of one blade element the relative air speed u at this element is needed. It can be divided into two parts: vertical component u_z and horizontal perpendicular to the blade u_r. The third component, parallel to the blade, is not relevant for calculated forces.

$$u_r = \omega_M r - \cos(\alpha)v_{wy} + \sin(\alpha)v_{wx} \tag{5.15}$$

$$u_z = v_{wz} \tag{5.16}$$

Figure 5.10 shows the relevant relationships of angles ϕ and forces. The pitch angle ϕ, which consists of collective pitch P_c and two cyclic pitches P_x and P_y[1],

[1] The signs of P_x and P_y depend on the rotating direction of the rotor.

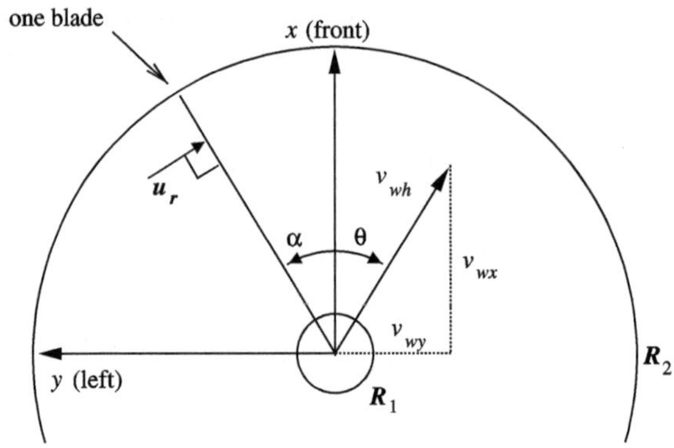

Fig. 5.9. Rotor disk for integration

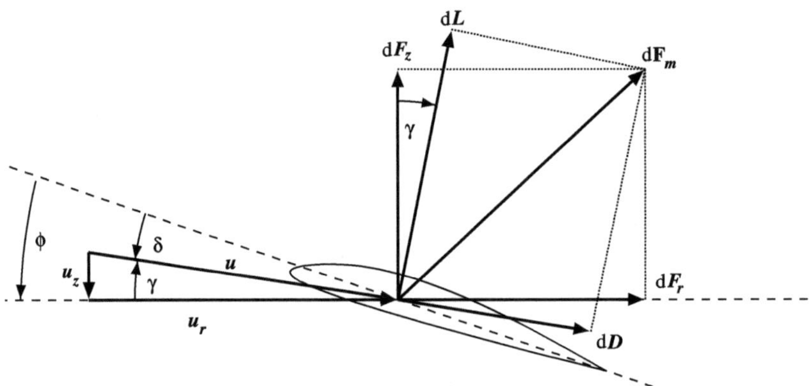

Fig. 5.10. Angles and air speed at one blade element

has to be computed as well as the direction of air flow γ, in order to obtain the angle of attack δ.

$$\phi = P_c \cos(\alpha) P_x + \sin(\alpha) P_y \tag{5.17}$$

$$\gamma = \operatorname{atan2}(u_z, u_r) \tag{5.18}$$

$$\delta = \phi + \gamma \tag{5.19}$$

According to [33, 22] lift dL (drag dD) is defined to be perpendicular (parallel) to the incoming air flow. Figure 5.11 depicts the relationship of lift and drag to angle of attack. With blade number N and chord width b, lift and drag are given by the following formulas. The slopes from Fig. 5.11 are approximated with the coefficients C_{l0}, C_{l1} and C_{d0}, C_{d2}. They can be determined by measurements

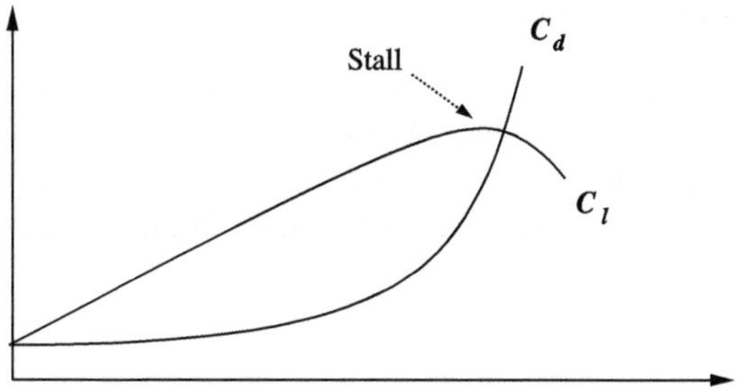

Fig. 5.11. Coefficients of lift and drag

during experiments. Since there are two unknown coefficients in each equation, at least two measurements at different operating points are required.

$$\mathrm{d}L = \frac{1}{2}\rho N b (C_{l0} + C_{l1}\delta) u^2 \tag{5.20}$$

$$\mathrm{d}D = \frac{1}{2}\rho N b (C_{d0} + C_{d2}\delta^2) u^2 \tag{5.21}$$

For the helicopter the forces $\mathrm{d}F_r$ and $\mathrm{d}F_z$ are relevant. They can be obtained using the inflow angle γ:

$$\mathrm{d}F_r = \mathrm{d}L \sin(\gamma) - \mathrm{d}D \cos(\gamma) \tag{5.22}$$

$$\mathrm{d}F_z = \mathrm{d}L \cos(\gamma) + \mathrm{d}D \sin(\gamma) \tag{5.23}$$

These values can now be used to obtain the vectors $\mathbf{F}_M$ and $\mathbf{M}_M$:

$$\mathrm{d}\mathbf{F}_M = \begin{pmatrix} -\mathrm{d}F_r \sin(\alpha) \\ \mathrm{d}F_r \cos(\alpha) \\ \mathrm{d}F_z \end{pmatrix} \tag{5.24}$$

$$\mathrm{d}\mathbf{M}_M = \begin{pmatrix} \mathrm{d}F_z r \sin(\alpha) \\ -\mathrm{d}F_z r \cos(\alpha) \\ \mathrm{d}F_r r \end{pmatrix} \tag{5.25}$$

Finally these vectors have to be integrated over the rotor disk. This has to be done numerically.

$$\mathbf{F}_M = \frac{1}{2\pi}\int_0^{2\pi}\int_{R_1}^{R_2} \mathrm{d}\mathbf{F}_M \mathrm{d}r \mathrm{d}\alpha \tag{5.26}$$

$$\mathbf{M}_M = \frac{1}{2\pi}\int_0^{2\pi}\int_{R_1}^{R_2} \mathrm{d}\mathbf{M}_M \mathrm{d}r \mathrm{d}\alpha \tag{5.27}$$

5.2.6 Simulation Results

This presented comparable simple model performs quite well in predicting flight behavior of an attached controller. At least with the controller as described in Sect. 5.3.1 below, the predicted trajectories are very similar to the actually flown ones from experiments. The trajectory of an individual distance flight is depicted in Fig. 5.12.

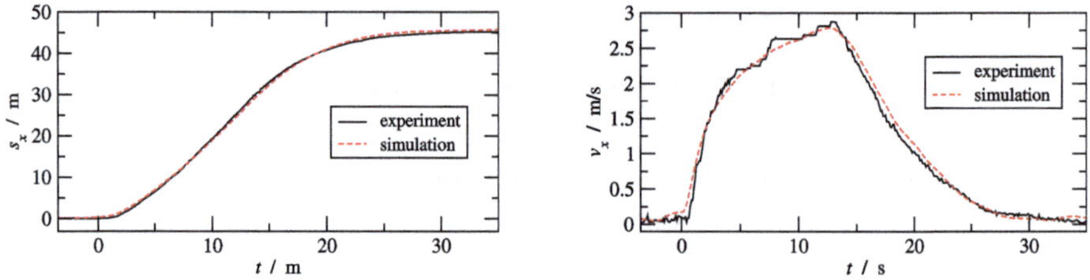

Fig. 5.12. Comparison of experiment and simulation

The position is predicted very accurately. The measurements of corresponding velocity show small deviation, mainly caused by wind gusts. For many applications such as planning or verifying a complete mission, this model is accurate enough.

But of course, there are several effects that cannot be described by this approach. One example is shown in Fig. 5.13.

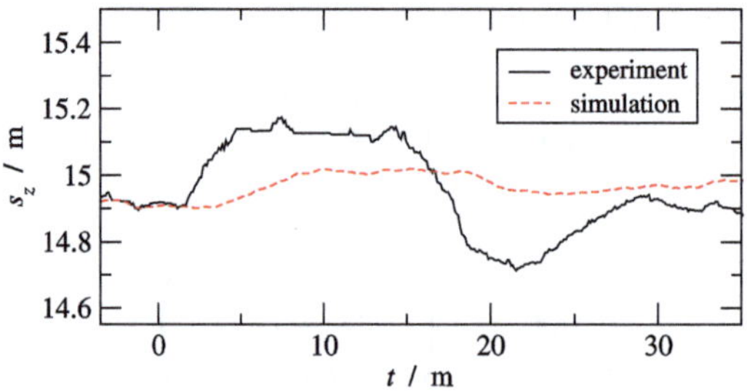

Fig. 5.13. Vertical position during a distance flight

Due to the so called Effective Translational Lift, which describes a gain in lifting efficiency of a horizontal moving rotor, at the beginning of the flight the helicopter climbs about 20 cm, before the controller takes care of that. During deceleration the effect is reversed, as the additional lift is lost at low velocities. Therefore, the presented model is not the final step in development. A discussion of a more elaborate model can be found in [8].

5.3 Control Techniques

5.3.1 MARVIN Controller

A helicopter has to be stabilized by a controller as it is inherently unstable by itself. Compared to other approaches such as [31, 45], this section presents an approach using a very simple model of a helicopter.

Model Based Elementary Controller Design

The complete control hierarchy of the helicopter control consists of several elementary controllers, each of which controlling a single scalar state variable. They are elementary linear controllers formally, but composed of two discrete parts. The "trajectory control" part realizes a first- or second-order low-pass filter on the command value for the closed control loop. The "error compensation" part corrects the controller's operating point in the manner of a first-order high-pass filter on the current model error.

This controller is stable by design and oscillation-free, its step response without overshooting, for any possible parameterization. This is an advantage over plain PID controllers that will always overshoot in case of just integrating plants.

Trajectory and System Model

Let x be the controlled variable and u the associated control variable. The first case to be considered is a double integrator with f as a constant of proportionality:

$$\ddot{x}(t) = \frac{1}{f} u(t) \tag{5.28}$$

When modelling a desired trajectory $x(t)$ to be followed by the system to reach a command value $x_r(t)$, it is obviously necessary to allow for two free parameters that can adopt to the current state $x(t_0)$ and $\dot{x}(t_0)$ given as initial condition at the current time t_0. $\ddot{x}(t_0)$ can be chosen by the controller according to (5.28).

In order to allow smooth convergence with at most one local extremum, a choice for the step response that suggests itself is

$$x(t) = x_r(t) + Ae^{-k_1 t} + Be^{-k_2 t} \tag{5.29}$$

with the constants $k_1, k_2 > 0$ adjusting the rate of convergence towards x_r and A, B representing the initial condition.

With equations (5.28) and (5.29) one can determine the basic control law, which reads

$$u(t) = f\ddot{x}^*(t) \tag{5.30}$$

$$\ddot{x}^*(t) := -k_1 k_2 (x(t) - x_r(t)) - (k_1 + k_2)(\dot{x}(t) - \dot{x}_r(t)) \tag{5.31}$$

$$\mathbf{C2} := -k_1 k_2 - s(k_1 + k_2) \tag{5.32}$$

Definition (5.31) can be interpreted as the desired acceleration, which is necessary in order to generate the intended trajectory according to equation (5.29). C2 represents the transfer function in frequency domain according to Laplace transformation with the corresponding complex variable s.

The coefficients have descriptive meanings: f is a system parameter which can be measured, and k_1 and k_2 define the convergence rate imposed by the controller and therefore define the half-life period of the exponential function.

The second case to be considered is a single integrator system:

$$\dot{x}(t) = \frac{1}{f}u(t) \tag{5.33}$$

In this case a single initial condition $x(t_0)$ has to be met, so that the trajectory only needs one free parameter A:

$$x(t) = x_r + Ae^{-k_0 t} \tag{5.34}$$

The corresponding control law is

$$u(t) = f\dot{x}^*(t) \tag{5.35}$$

$$\dot{x}^*(t) := -k_0(x(t) - x_r) \tag{5.36}$$

$$\mathrm{C1} := -k_0 \tag{5.37}$$

Transient and Stationary Phases

Within this control approach the command value x_r is not assumed to prescribe a desired state trajectory in the form of a continuous guiding signal. The command value shall always indicate some "final" destination, such that x_r is subject to occasional and substantial jumps and the corresponding desired response of the system is fully defined by trajectories (5.29) or (5.34) and taken care of by the controller.

Especially in cases of considerable jumps of the command value, it is reasonable to distinguish phases of keeping a position (position control) on the one hand and travelling (velocity control) on the other.

For this, a favorable approach is to use the velocity controller's desired acceleration (5.36) for travelling in either positive or negative direction at a desired magnitude of velocity $v_r > 0$ as the clipping boundaries for the use of the position controller (5.31):

$$a^- := -k_0(\dot{x}(t) - (-v_r))$$

$$a^+ := -k_0(\dot{x}(t) - (+v_r))$$

$$a := -k_1k_2(x(t) - x_r(t)) - (k_1 + k_2)(\dot{x}(t) - \dot{x}_r(t)) \tag{5.38}$$

$$\ddot{x}^*(t) := \begin{cases} a^- & \text{if} \quad a \leq a^- \\ a & \text{if} \quad a^- < a < a^+ \\ a^+ & \text{if} \quad a \geq a^+. \end{cases}$$

Phase switching according to (5.38) guarantees a continuous output signal. Additionally, clipping due to the velocity controller's outputs provides a convenient limitation of the possible resulting control outputs, only due to v_r and k_0.

Operating Point and Model Error

In order to eliminate steady state errors, it is necessary to provide some means of adjusting a controller's operating point. The reason for the necessity of such compensation is that real systems differ from the ideal equations (5.28) or (5.33). Any offset of the linear equation or external disturbance is not handled. Using the double integrator case as an example, (5.28) changes for real systems with the model error $z(t)$ into

$$\ddot{x}(t) = \frac{1}{f}u(t) - z(t) \tag{5.39}$$

The controller as in [30] is extended by an estimator $z_{est}(t)$ of the model error, which contributes to the controller output:

$$u(t) = f(\ddot{x}^*(t) + z_{est}(t)) \tag{5.40}$$

This estimator $z_{est}(t)$ is gradually adjusted at a selectable rate a by integrating the deviation between the current *desired* $\ddot{x}^*(t)$ and the current actual *observed* $\ddot{x}(t)$:

$$\dot{z}_{est}(t) = a(\ddot{x}^*(t) - \ddot{x}(t)) \tag{5.41}$$

Summary of Elementary Controller

Figure 5.14 sums up the composition of one double-integrator type elementary controller as constructed in the course of this section. The trajectory control and error compensation parts are indicated as more abstract building blocks C2 and EC, respectively.

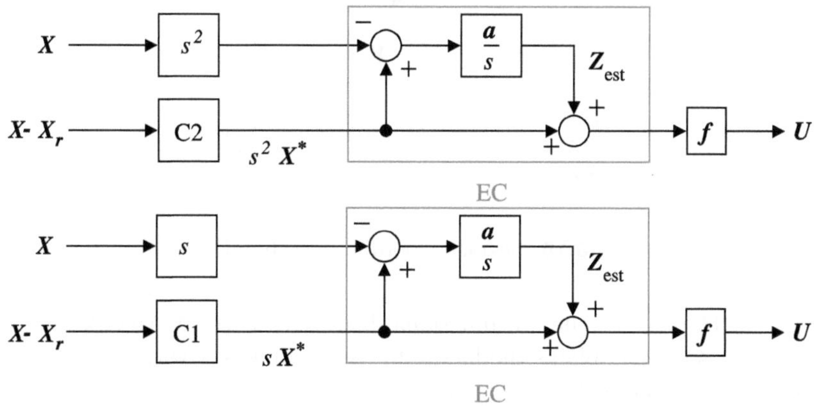

Fig. 5.14. Block diagrams for complete elementary controllers (trajectory and operating point). Capital letters denote Laplace transformed signals.

A single-integrator type elementary controller as constructed above is also shown. Please note that in phase switching according to equation (5.38), the "level of derivatives" is generally raised by one, so that one EC block would be shared by the two alternative trajectory control blocks C1 and C2.

These two figures reveal the differences compared to PID controllers. The latter only accept inputs of the control deviation $x - x_r$ but in Fig. 5.14 x is used directly. Thus, the presented elementary controllers use two input signals (x and $x - x_r$), whereas PID controllers only use $x - x_r$. The result is a trajectory without overshooting, which is not possible with a plain PID controller in case of single or double integrator plants. Apart from that the coefficients f, a and k_i have descriptive meanings, that are easy to determine.

Control Hierarchy

The helicopter control hierarchy uses a dedicated coordinate system. The segment coordinate system (SCS) is defined according to the current desired course segment, which is always a straight line through two points, $\mathbf{P}_{from}$ and $\mathbf{P}_{to}$ (see Fig. 5.15). x is parallel to the course line, y is always horizontal and z to the upper BCS half-space. Below the coordinate system of a vector is denoted by a single prefixed superscript letter (e.g. ${}^{\mathrm{B}}\mathbf{s}$ for the position given in BCS).

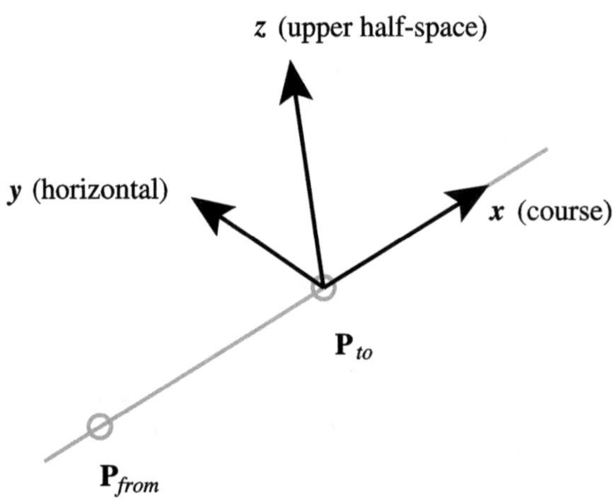

Fig. 5.15. Segment coordinate system (SCS)

Figure 5.16 presents the full functional hierarchy of the MARVIN flight controller. This section explains the hierarchy and connects it with the formal considerations from the previous sections.

The complete hierarchy is executed at a control frequency of $T^{-1} = 40\,\text{Hz}$. Flight experiments at $T^{-1} = 20\,\text{Hz}$ also worked without visible degradation of control performance. Even $T^{-1} = 10\,\text{Hz}$ worked flawlessly in simulation. Higher frequencies than 40 Hz will not lead to better control results because of two main reasons. The main rotor rotates at approximately 20 Hz and consists of two blades, which means that averaging its forces and torques is approximately correct at 40 Hz. The second reason is the limited update rate the servos are capable of. Conventional analog servo devices accept positioning signals at rates of approximately 50 Hz, so that higher commands could not be sent to the actuators.

In the diagram, boxes labelled C1/2, C1, and EC refer to the corresponding controller and error compensation modules as introduced in Fig. 5.14. C1/2 specifically refers to the phase-switching controller detailed in equation (5.38). All blocks with dashed outline constitute coordinate transformations with generally no impact on system dynamics. To avoid new identifiers representing the individual signals in Laplace domain, the time domain identifiers are used in the diagram.

In the top left region, the SCS-position and -velocity are fed through three phase-switching controllers C1/2 based. The underlying physical relationship is Newton's law, which is indeed a simple linear double integrator as required for the elementary controllers. As ${}^{\mathrm{S}}\mathbf{s}$ already incorporates the reference position $\mathbf{P}_{to}$ (it is given in SCS), it can be used directly as input. The result is the desired SCS acceleration vector ${}^{\mathrm{S}}\mathbf{a}^*$. It is important to note that flight path segments are usually given such that only the SCS-x controller will apply velocity control at greater distances from $\mathbf{P}_{to}$, while the y and z controllers just have to keep the helicopter laterally on the course line using position control.

In the next step, the desired acceleration is just transformed into BCS orientation, yielding ${}^{\mathrm{B}}\mathbf{a}^*$. This means that there is a coordinate transformation applied to the desired response variable *between* trajectory control and error compensation stages.

The BCS-horizontal components of ${}^{\mathrm{B}}\mathbf{a}^*$ are then fed through error compensation blocks in order to incorporate operating point biases. These biases are mainly intended to compensate for wind effects, which is why the BCS is the only "correct" coordinate system to anchor these biases in. The results are labelled ${}^{\mathrm{B}}a_x^{*'}$, ${}^{\mathrm{B}}a_y^{*'}$ indicating that they express "virtual" accelerations with a descriptive meaning of required "force" to be *actually applied*, but scaled as acceleration (force and acceleration are coupled via the mass m as usual).

The BCS-vertical component ${}^{\mathrm{B}}a_z^*$ needs to be handled differently. The vertical force offset is mainly constant (gravity g) and model errors tend to result in changes of the *actually applied* force $\mathbf{F}_{mt}$ of the rotors, not in changes of the *required* force as in wind compensation. The changes in the magnitude of $\mathbf{F}_{mt}$ are aerodynamically caused by induced velocity effects. Therefore, ${}^{\mathrm{B}}a_z^{*'}$ is obtained by just adding g, and model error compensation will effectively be applied in the course of collective pitch control. In other words, the operating point for ${}^{\mathrm{B}}\mathbf{a}^{*'}$ is adjusted in non-orthogonal coordinates defined by BCS-x, BCS-y, and $\mathbf{F}_{mt}$.

Before discussing collective pitch, it is reasonable to discuss cyclic pitch control first. The cyclic pitches ϕ_x and ϕ_y need to align the direction of the applied force $\mathbf{F}_{mt}$ with the direction of the desired acceleration $\mathbf{a}^{*'}$ as requested by the SCS navigation controllers explained above. As attitude is controlled by cyclic pitches,

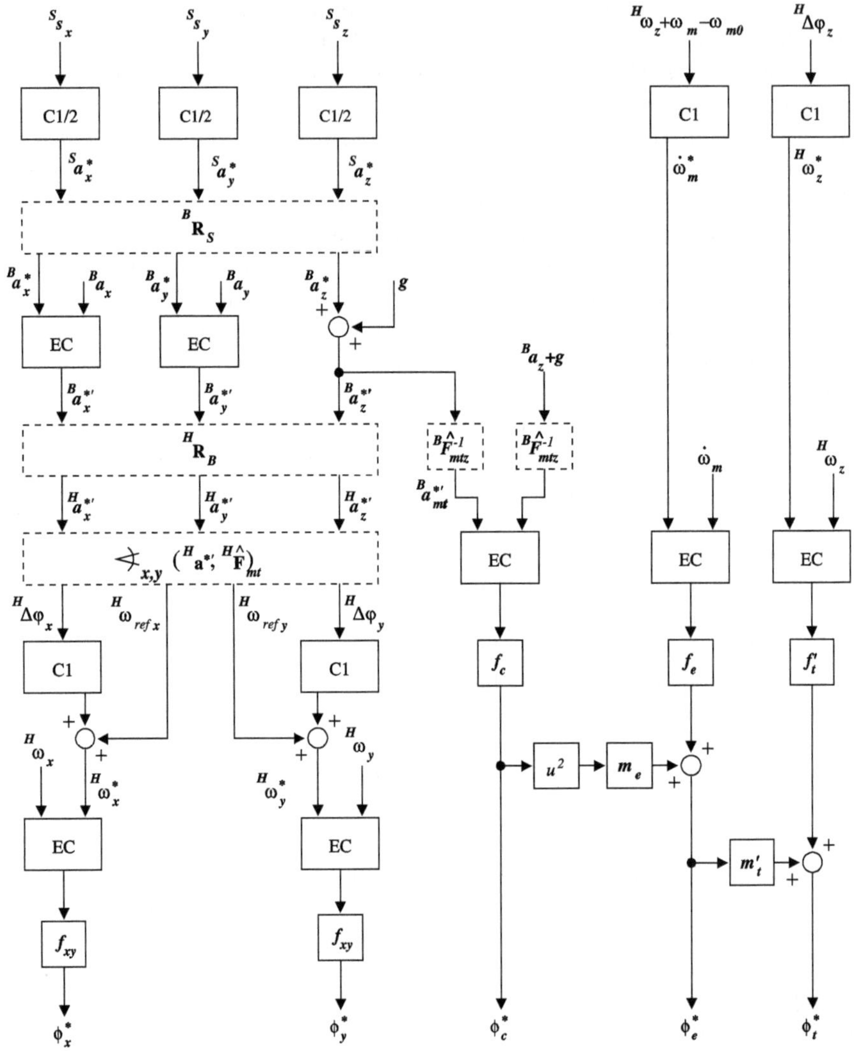

Fig. 5.16. Full hierarchy of the MARVIN flight controller

that operate in HCS, $\mathbf{a}^{*'}$ is transformed into HCS first. The next transformation block calculates a rotation vector ${}^{\mathrm{H}}\Delta\varphi = ({}^{\mathrm{H}}\Delta\varphi_x \quad {}^{\mathrm{H}}\Delta\varphi_y \quad 0)^T$, that expresses the required HCS-x, rotation between $\mathbf{F}_{mt}$ and $\mathbf{a}^{*'}$.

${}^{\mathrm{H}}\Delta\varphi$ constitutes the control error for the cyclic pitch controllers, the two C1 blocks in the next stage. As ${}^{\mathrm{H}}\Delta\varphi$ is a two-dimensional controlled variable, it is necessary to validate its dynamics resulting from the controllers' operation according to equation (5.36). Both C1 controllers use the same k_0. Therefore, it can be seen that the resulting rotation rate ${}^{\mathrm{H}}\omega$ is always parallel to ${}^{\mathrm{H}}\Delta\varphi$. Thus, ${}^{\mathrm{H}}\Delta\varphi(t)$ as a vector signal does indeed follow the convergence law (5.34).

The above statement is only valid if the reference direction of $\mathbf{a}^{*'}$ does not change. Therefore, the previous transformation block also calculates the change rate of this reference direction ${}^{\mathrm{H}}\omega_{ref}$. As this change rate is directly proportional to the required cyclic pitch, it can just be mixed into the trajectory controllers' ${}^{\mathrm{H}}\omega^*$ outputs, as shown in the diagram. The two error compensation (EC) blocks and the final transformation into controller outputs ϕ_x^* and ϕ_y^* via multiplication by f_{xy} conclude cyclic pitch control[2].

While the orientation of the rotors' force (denoted by the unit vector $\hat{\mathbf{F}}_{mt}$) is subject to cyclic pitch control, collective pitch control addresses the magnitude of $|\mathbf{F}_{mt}|$. Technically, ${}^{\mathrm{B}}F_{mtz}$ shall equal $m{}^{\mathrm{B}}a_z^{*'}$. Formally, this means

$$
\begin{aligned}
{}^{\mathrm{B}}F_{mtz} &= |\mathbf{F}_{mt}|{}^{\mathrm{B}}\hat{F}_{mtz} = m{}^{\mathrm{B}}a_z^{*'} \\
a_{mt}^{*'} &:= \frac{|\mathbf{F}_{mt}|}{m} = \frac{{}^{\mathrm{B}}a_z^{*'}}{{}^{\mathrm{B}}\hat{F}_{mtz}}
\end{aligned}
\tag{5.42}
$$

In the diagram, this is depicted as the desired input to the EC block of collective pitch ϕ_c^*. The observed input to this error compensation block, ${}^{\mathrm{B}}a_z + g$, needs to undergo the same inverse projection to refer to the $\mathbf{F}_{mt}$ direction, so that it can reasonably be compared to the desired value within the EC block. The final multiplication by the model parameter f_c concludes collective pitch control.

Rpm (throttle ϕ_e^*) control is a very straightforward C1 and EC single-integrator type. It uses a command value ω_{m0} that remains constant during flight. Additionally, after applying the model factor f_e, there is a mixer that adds $m_e\phi_c^{*2}$ to the throttle output ϕ_e^*. This compensates for the effect of collective pitch changes on the torque of the main rotor. Using a collective-pitch-to-throttle mixer like this is also common in remotely controlled model helicopters.

Figure 5.16 depicts the tail rotor control for the case of a hardware gyroscope module in the loop. Thus, structurally, the controller for the tail rotor signal ϕ_t^* (which actually goes to the commercial gyroscope module) is identical to throttle control. There is also a mixer, this time a linear throttle-to-tail-rotor mixer with the factor m_t. This is also well known from remote control helicopters.

If the hardware gyroscope module is missing, it is possible to substitute the tail control with the structure of a C2 controller according to Fig. 5.14. But since such a hardware controller performs very well and first of all can operate with a very small latency, it is recommendable to use one if possible.

The source of the heading error ${}^{\mathrm{H}}\Delta\varphi_z$ for tail rotor control still needs to be explained. During flight, the controller tries to align the helicopter's heading with the horizontal projection of the current course segment. This basically means that the helicopter always flies "forward". Formally, the desired heading (in the BCS-xy-plane) is transformed into the HCS, projected into the HCS-xy-plane, and its angle with HCS-x then used as the control error ${}^{\mathrm{H}}\Delta\varphi_z$.

2 One issue left out here is that the observed acceleration ${}^{\mathrm{B}}a_x$, ${}^{\mathrm{B}}a_y$ as used by the desired force EC blocks in the navigation ("outer") control loop will be adversely affected by any transient control error in the cyclic pitch ("inner") control loop. The implementation of the controller does take care of that, thus minimizing any unwanted cross-talk between different control loops.

Experimental Results

The previously described controller is used by MARVIN for real flight experiments. In conjunction with a ground station it is capable of flying complete missions. Figure 5.17 depicts the trajectory of such a flight experiment.

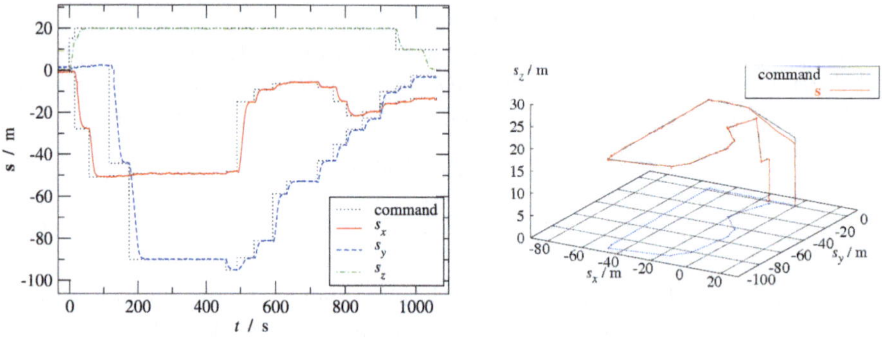

Fig. 5.17. A complete mission flight, depicted with individual coordinates and as 3D projection

The left figure shows the three individual BCS coordinates of the trajectory that follow the dotted command values according to the exponential approach according to equation (5.29). On the right, the same flight is depicted in 3D.

The presented control approach is able to successful control a helicopter and follow the given trajectories accurately. For control purposes, there is no need for a more elaborated model, as long as no extreme manoeuvres are intended.

5.3.2 Additional Control Techniques for MARVIN

As it was stated in the first section of this chapter, several control methods have been applied in literature for control of UAVs, just in simulation or with real vehicles. Obviously, the applicable control methods highly depend on the capabilities of the on-board hardware. Concerning MARVIN a series of control techniques have been implemented and evaluated, in simulation and/or with real flights, including the hierarchy of elementary linear controllers exposed in Sect. 5.3.1 (it will be referred as HELC in this section), LQR, H_∞, nonlinear control and fuzzy control. Each one has its own advantages and drawbacks when performance, robustness, design process, and/or practical applicability within the limited MARVIN on-board fixed point micro-controller are considered. Therefore, a trade-off of advantages and drawbacks has to be taken into account.

From the above considerations, it can be said that the HELC structure and fuzzy control are the simplest approaches, showing a good performance and robustness with low computational cost. More complex control laws, with an improved performance, can be considered but the main problem is its practical application within the MARVIN on-board micro-controller.

The HELC structure, described in previous sections, is probably the simplest approach. This controller shows a good performance and can be used for both hovering and trajectory tracking exhibiting a good robustness. Moreover, even if a model of the helicopter is not needed for tuning the controller, it can take advantage of the available model (described in Sect. 5.2 of this Chapter), as it was really done with MARVIN, to adjust the controller parameters in simulation to obtain the desired performance. Another advantage is its very low computational cost making it possible to implement it within the fixed point micro-controller. On the other hand, the main drawback is that low gains have to be used due to saturations and large errors, although this is partially solved by using different controllers for position and velocity control.

Concerning the fuzzy controllers, their main advantages are, on the one hand, that they can be used as nonlinear controllers considering different operation points with their corresponding control laws to improve performance, to deal with saturations or, even, to induce nonlinear behaviors on the helicopter flight. In this way, for instance, within the fuzzy controller it is possible to apply different control laws according to the operation conditions, improving the overall controller performance. For instance, in the HELC control scheme, the parameters of each controller (proportional, derivative and integral action) are fixed for all the operation range. However, within a fuzzy controller, these parameters can be changed according to the operation conditions to provide the most effective control action without saturating, for instance. On the other hand, they can be implemented with low computational cost, similar to the required by the HELC structure.

Fuzzy Controller for MARVIN

Before continuing, it has to be pointed out that, from a practical point of view, performance of the current HELC structure is good enough to execute MARVIN flights successfully. Therefore, improving performance was not required as a goal of the fuzzy control design. Moreover, the HELC structure controller can be seen as a pseudo-fuzzy controller. On the one hand, when the destination point is close to the origin point, *hold* phase, a position control is applied, because the resultant inputs will not imply high speeds. However, if the distance to the destination point is bigger (threshold value is 10 m), the *approach* phase, input action of position control would lead to high velocities, which in its turn would cause lack of stability of the helicopter. Therefore, a velocity control is used in this *approach* phase. Both controllers are combined by selecting the minimum control action each time instant.

Thus, the use of fuzzy control within MARVIN was focused on testing MARVIN model under nonlinear effects by comparing the results computed in simulation with those obtained in field experiments, rather than on improving control performance. Namely, fuzzy control has been used to induce nonlinear behaviors (including multiple equilibriums, limit cycles, etc) in order to validate the

MARVIN model with experimental flights in the face of such a nonlinear behaviors. To that end, the proposed fuzzy control law (a Takagi-Sugeno type) is a variation of the HELC structure in the vertical dynamics.

Takagi-Sugeno fuzzy controllers are composed, in general, by M rules of the form

$$R_i : \text{IF } x_1 \text{ IS } F_i^1 \text{ AND...AND } x_n \text{ IS } F_i^n \text{ THEN } u_i = f_i(x_i, ..., x_n), \text{ for } i = 1, ..., M$$

where F_i^j are fuzzy sets describing the variables x_j, and $f_j(...)$ are linear, affine or nonlinear functions of a set of variables. Taking all the rules in mind, the following nonlinear control law is obtained

$$u = \sum_{i=1}^{M} \frac{\prod_{j=1}^{n} \mu_{F_i^j}(x_j) f_i(x_1, ..., x_n)}{\sum_{i=1}^{m} \prod_{j=1}^{n} \mu_{F_i^j}(x_j)}$$

where functions $\mu_{F_i^j}(x_j)$ represent the membership degree of the variable x_j to the fuzzy set F_i^j. In the MARVIN case, the proposed Takagi-Sugeno fuzzy controller is composed of a set of rules such as:

$$\text{IF Altitude\_Error IS Medium THEN } pc_i = k_{pi} e_z + k_{di} \dot{e}_z + k_{ii} \int e_z dt$$

where different control gains (k_{xi}) can be used in each rule. The resulting control action (the collective input of the main rotor, *pc*) is obtained as a weighted average of the output of each rule (pc_i). In the same way, fuzzy rules are used for controlling the tail rotor. Keeping in mind its practical application within MARVIN, a similar effect, but with a minimal modification of the code of the controller, can be obtained by using a fuzzy system to select the value of the control parameter k_2 in the type 2 control law of the HELC scheme, depending on the altitude error.

The altitude fuzzy controller was designed in such a way that the system exhibit the behavior shown in Fig. 5.18 corresponding to the "altitude_error vs. vertical_velocity" phase plane. As it can be observed, the system presents five equilibrium points, which are surrounded by a stable limit cycle. The origin is the only one stable equilibrium, but it is only locally stable; there are also two unstable equilibrium points and two saddle points between the stable and the unstable equilibria.

In this way, if the helicopter starts close to the target altitude, i.e., with a small altitude error, it will be able to reach the origin (it achieves null altitude error). However, if the error is a little bit larger (see dashed line in Fig. 5.18), or a disturbance brings the helicopter far from the target point, it can be out of the attraction basin of the origin and then the helicopter will tend to go away from the origin. Notice that the existence of the stable limit cycle prevents the helicopter to become totally unstable. Instead of this, MARVIN will exhibit a

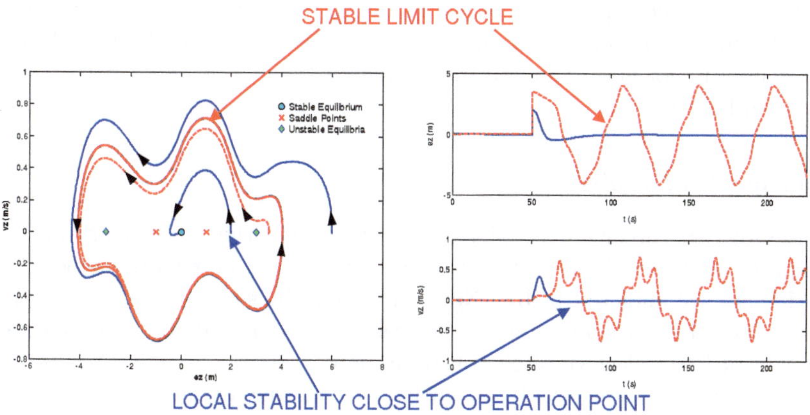

Fig. 5.18. "Altitude_error vs. vertical_velocity" phase plane and temporal response from different initial altitude error

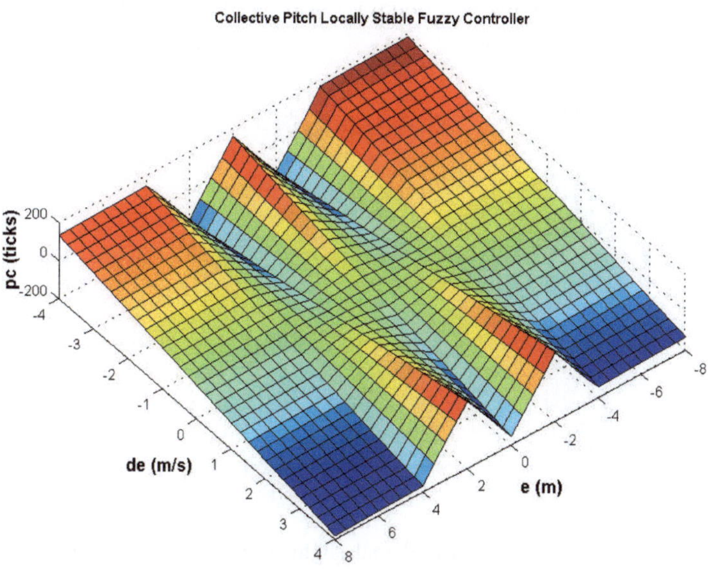

Fig. 5.19. Nonlinear control surface of the locally stable altitude fuzzy controller

permanent stable oscillation. If the error is even larger, the stable limit cycle will attract the helicopter preventing again from instability (see Fig. 5.18). Figure 5.19 shows the nonlinear control surface of the locally stable fuzzy controller.

Clearly, from the point of view of model validation, testing a controller like this one on the real helicopter has several advantages. On the one hand, it makes it possible to compare the response of the real helicopter and the model in the face of non globally stable controllers. On the other hand, it makes it also possible to test the capabilities of the model to reflect the nonlinear behavior of MARVIN by comparing the predicted oscillations and the real ones. Moreover, if the model is not good enough, the predicted limit cycle could not exist with the actual helicopter.

Experimental Results

A series of experiments with MARVIN flying with this fuzzy controller have been performed to validate the aforementioned simulation results. Some of these experiments are depicted below. Notice that for the sake of safety, low vertical velocity should be provided during the oscillations in real flight. This is achieved by means of low collective pitch control action.

Simulating helicopter velocity evolution with the fuzzy controller and an initial altitude error of 5 meters, and comparing with the real MARVIN behaviour, it can be observed that MARVIN evolves faster than the model. Figure 5.20, shows the corresponding "altitude error vs. z velocity" phase plane. However, it can be observed that both, simulated (dark grey) and real experiments (light grey), present an equivalent qualitative behavior starting far away from the operation point, i.e., a stable limit cycle.

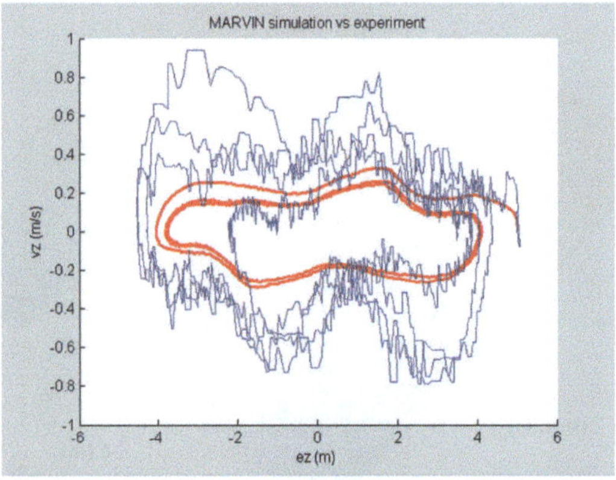

Fig. 5.20. Simulated (dark grey) and real (light grey) MARVIN phase plane for an initial altitude error of 5 meters

In order to analyze the behaviour closer to the origin a new experiment was performed starting at 2.66 meters from the equilibrium point. As can be observed in Fig. 5.21, when MARVIN starts at 2.66 m from the origin it tends to the stable equilibrium. However, that stability is only local. A perturbation appeared during the real flight bringing MARVIN out of the attraction basin of the origin, so it evolved to the same stable limit cycle than in the previous experiment with an initial error of 5 meters.

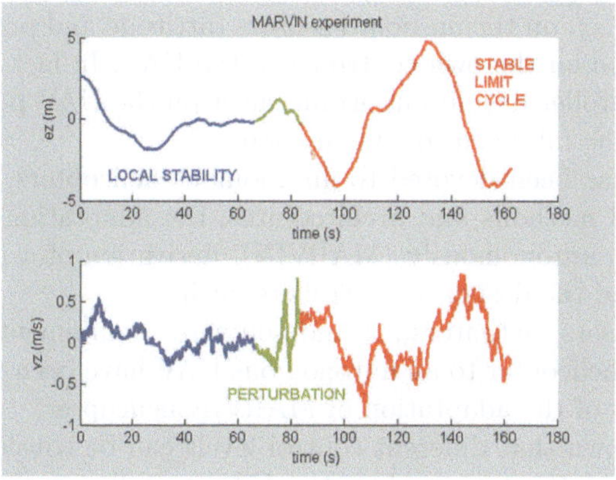

Fig. 5.21. Real experiment with MARVIN: it starts 2.66 m far from the origin and it tends to the equilibrium; a perturbation brings MARVIN to the stable limit cycle

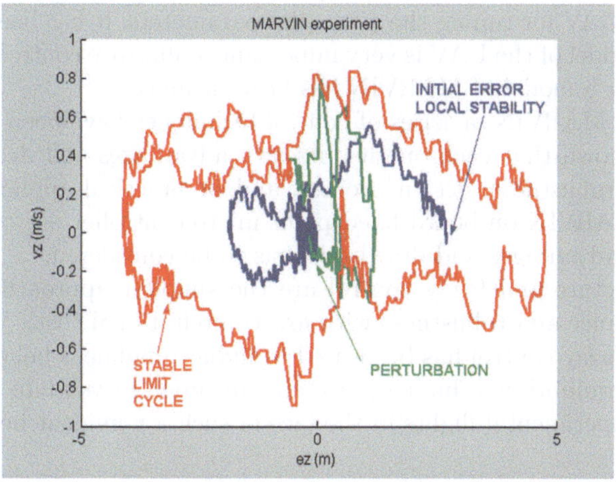

Fig. 5.22. Phase plane for MARVIN real experiment with initial altitude error of 2.66 meters. A perturbation brings MARVIN out of the attraction basin of the origin.

The final conclusion is that the nonlinear qualitative behaviour of the MARVIN model with the fuzzy controller is quite similar to that of the real MARVIN with this fuzzy controller. However, some differences on the quantitative velocity behaviour have been found, probably due to non modelled effects of induced velocity.

5.4 Conclusions

Performance of a multi UAV system depends on the capabilities of each UAV, and such capabilities depend, among others, on the type of UAV (helicopter, airship, airplane, ...), on the on-board sensors (attitude and position estimation, autonomy, ...), and on the own controller of the UAV. In fact, the performance of each UAV controller has an important effect on the UAV performance which strongly impact the fulfilment of the mission.

This chapter has been devoted to autonomous helicopters, including a brief survey on control methods and architectures, the adaptation of a model helicopter to perform autonomously (MARVIN), derivation of a physical model of the helicopter, and the design of controllers for it.

Hardware (sensors, actuators,...) and software requirements for changing a remotely piloted helicopter to an autonomous UAV have been illustrated in this chapter by means of the adaptation of MARVIN helicopter.

It has been shown that different control levels can be considered: Trajectory tracking control, at the top, enables the UAV to follow a trajectory which is usually given by means of way-points; Low level control is required to stabilize the UAV in spite of perturbations. Usually, the low level control is in charge of stabilizing the attitude, and positions for hovering purposes.

Most of the applied control techniques are based on a model (nonlinear or linear) of the UAV for tuning the controller parameters. It has been pointed out that a good model of the UAV is very important to improve control performance. In this chapter a model of MARVIN has been included.

Concerning MARVIN, a series of control techniques have been implemented. It has been shown that each one has its own advantages and drawbacks when performance, robustness, design process, and/or practical applicability within the limited MARVIN on-board fixed point micro-controller are considered. So, a trade-off of advantages and drawbacks has to be considered.

HELC structure and fuzzy control are the simplest approaches, showing a good performance and robustness with low computational cost.

Moreover, fuzzy control has been used to induce nonlinear behaviors (including multiple equilibriums, limit cycles, etc) in order to validate the MARVIN model with experimental flights in the face of such a nonlinear behaviors.

References

1. ACTeurope. Germany. `http://www.acteurope.de`.
2. Aero-Tec. CB-5000, Germany. `http://www.aero-tec-helicopter.de`.
3. Höft & Wessel AG. Hw8615. `http://www.hoeft-wessel.com`.
4. G. Buskey, G. Wyeth, and J. Roberts. Autonomous helicopter hover using an artificial neural network. In *Proceedings of the IEEE International Conference on Robotics & Automation*, pages 1635–1640, Seoul, Korea, May 2001.
5. C. Cavalcante, J. Cardoso, J.G. Ramos, and O.R. Nerves. Design and tuning of a helicopter fuzzy controller. In *Proceedings of IEEE International Conference on Fuzzy Systems*, volume 3, pages 1549–1554, 1995.
6. KOMATSU ZENOAH CO. Japan. `http://www.zenoah.net`.
7. COMETS consortium. Real-time coordination and control of multiple heterogeneous unmanned aerial vehicles. IST 34304, 5th Framework Program, 2002. `http://www.comets-uavs.org`.

8. C. Deeg. *Modeling, Simulation, and Implementation of an Autonomously Flying Robot.* PhD thesis, Technische Universität Berlin, dissertation.de - Verlag im Internet GmbH, July 2006.
9. C. Deeg, M. Musial, and G. Hommel. Control and simulation of an autonomously flying model helicopter. In *Proceedings of the 5th IFAC Symposium on Intelligent Autonomous Vehicles*, Lisboa, Portugal, 2004.
10. P. Doherty, G. Granlund, K. Kuchcinski, E. Sandewall, K. Nordberg, E. Skarman, and J. Wiklund. The WITAS unmanned aerial vehicle project. In *Proceedings of the 14th European Conference on Artificial Intelligence*, pages 747–755, Berlin, Germany, 2000.
11. G. Done and D. Balmford. *Bramwell's Helicopter Dynamics.* Butterworth Heinemann, second edition, April 2001.
12. Frenzel + Berg Elektronik. Germany. `http://www.frenzel-berg.de`.
13. A.H. Fagg, M.A. Lewis, J.F. Montgomery, and G.A. Bekey. The USC autonomous flying vehicle: An experiment in real-time behaviour-based control. In *Proceedings of the IEEE/RSJ International Conference on Intelligent Robots and Systems*, pages 1173–1180, July 2003.
14. I. Fantoni and R. Lozano. *Non-linear Control for Underactuated Mechanical Systems*, chapter 13: Helicopter on a platform. 2002.
15. Association for Unmanned Vehicles International. International Aerial Robotics Competition: The Robotics Competition of the Millennium. `http://avdil.gtri.gatech.edu/AUVS/`.
16. V. Gavrilets, E. Frazzoli, B. Mettler, M. Piedmonte, and E. Feron. Aggressive maneuvering of small autonomous helicopters: A human-centered approach. *The International Journal of Robotics Research*, 20(10):795–807, 2001.
17. V. Gavrilets, B. Mettler, and E. Feron. Non-linear model for a small-size acrobatic helicopter. In *Proceedings of the AIAA Guidance, Navigation and Control Conference*, Montreal, Canada, August 2001.
18. Novatel Inc. Canada. `http://www.novatel.ca`.
19. SensComp Inc. USA. `http://www.senscomp.com/600smartsensor.htm`.
20. A. Isidori, L. Marconi, and A. Serrani. Robust nonlinear motion control of a helicopter. In *Proceedings of the 40th IEEE Conference on Decision and Control*, pages 4586–4591, Orlando, Florida, USA, December 2001.
21. E.N. Johnson and S. K. Kannan. Adaptive flight control for an autonomous unmanned helicopter. In *Proceedings of the AIAA Guidance, Navigation and Control Conference*, number AIAA-2002-4439, Monterey, CA, August 2002.
22. Wayne Johnson. *Helicopter Theory.* Dover Publications, Inc., New York, USA, 1994.
23. B. Kadmiry, P. Bergsten, and D. Driankov. Autonomous helicopter using fuzzy-gain scheduling. In *Proceedings of the IEEE International Conference on Robotics & Automation*, volume 3, pages 2980–2985, Seoul, Korea, May 2001.
24. Graupner GmbH & Co. KG. Germany. `http://www.graupner.de`.
25. H.J. Kim and D.H. Shim. A flight control system for aerial robots: algorithms and experiments. *Control Engineering Practice*, 11:1351–1515, 2003.
26. S.K. Kim and D.M. Tilbury. Mathematical modeling and experimental identification of a model helicopter. *Journal of Robotic Systems*, 21(3):95–116, 2004.
27. Kokam. Korea. `http://www.kokam.co.kr/english/index.html`.
28. K. Kondak, C. Deeg, G. Hommel, M. Musial, and V. Remuß. Mechanical model and control of an autonomous small size helicopter with a stiff main rotor. In *Proceedings of the IEEE/RSJ International Conference on Intelligent Robots and Systems*, pages 2980–2985, 2004.

29. T.J. Koo, F. Hoffman, H. Shim, B. Sinopoli, and S. Sastry. Hybrid control of model helicopters. In *Proceedings of the IFAC Workshop on Motion Control*, pages 285–290, Grenoble, France, 1998.
30. T.J. Koo, Y. Ma, and S. Sastry. Nonlinear control of a helicopter based unmanned aerial vehicle model. *IEEE Transactions on Control Systems Technology*, 2001.
31. T.J. Koo and S. Sastry. Output tracking control design of a helicopter model based on approximate linearization. In *Proceedings of the 37th IEEE Conference on Decision and Control*, pages 3635–3640, Tampa, Florida, USA, 1998.
32. Lambda. Europe. `http://www.lambdaeurope.com`.
33. J. Gordon Leishman. *Principles of Helicopter Aerodynamics.* Cambridge University Press, 2000.
34. CompuLab Ltd. Israel. `http://www.compulab.co.il`.
35. M.F. Griffin M. Sugeno and A. Bastian. Fuzzy hierarchical control of an unmanned helicopter. In *Proceedings of the 17th IFSA World Congress*, pages 179–182, 1993.
36. E. Madelung. *Die mathematischen Hilfsmittel des Physikers.* Springer, Heidelberg, 7 edition, 1964.
37. D.Y. Maharaj. *The application of non-linear control theory to robust behaviour-based control.* PhD thesis, Dept of Aeronautics, Imperial College of Science, Technology and Medicine, 1994.
38. M. Mettler, M. B. Tischler, and T. Kanade. System identification modeling of a small-scale unmanned rotorcraft for flight control design. *American Helicopter Society Journal*, 2002.
39. Microstrain. GX1, USA. `http://www.microstrain.com`.
40. J.F. Montgomery and G.A. Bekey. Learning helicopter control through "teaching by showing". In *Proceedings of the 37th IEEE Conference on Decision and Control*, December 1998.
41. J.F. Montgomery, A.H. Fagg, and G.A. Bekey. The USC AFV-i: A behavior-based entry in the 1994 international aerial robotics competition. *IEEE Expert*, 10(2):16–22, 1995.
42. M. Musial, U.W. Brandenburg, and G. Hommel. MARVIN – technische universität berlin's flying robot for the IARC Millennial Event. In *Proc. Symposium of the Association for Unmanned Vehicle Systems 2000*, Orlando, Florida, USA, 2000.
43. M. Musial, C. Deeg, V. Remuß, and G. Hommel. Orientation sensing for helicopter UAVs under strict resource constraints. In *Proceedings of the First European Micro Air Vehicle Conference (EMAV)*, pages 13–14, Braunschweig (Germany), July 2004.
44. Hamamatsu Photonics. Japan. `http://www.hamamatsu.com`.
45. Rüdiger Reichow. *Ein robuster Hubschrauber–Flugregler.* PhD thesis, Technische Universität Carolo-Wilhelmina zu Braunschweig, Fakultät für Maschinenbau und Elektrotechnik, 29, September 1995.
46. V. Remuß and M. Musial. Communication system for cooperative mobile robots using ad-hoc networks. In *Proceedings of the 5th IFAC Symposium on Intelligent Autonomous Vehicles*, Lisbon, Portugal, 2004. Elsevier Science, ISBN 008-044237-4.
47. V. Remuß, M. Musial, and U.W. Brandenburg. BBCS – robust communication system for distributed systems. In *Proc. IEEE International Workshop on Safety, Security and Rescue Robotics (SSRR).* ISBN 3-8167-6556-4, May 2004.
48. H. Shim, T.J. Koo, F. Hoffman, and S. Sastry. A comprehensive study of control design of an autonomous helicopter. In *Proceedings of the 37th IEEE Conference on Decision and Control*, pages 3653–3658, Tampa, Florida, USA, July 1998.

6

Airship Control

Leonardo Solaque[1,*] and Simon Lacroix[2,*]

[1] De los Andes University, Carrera 1ra No. 18A-10, Bogotá (Colombia)
l-solaqu@uniandes.edu.co

[2] LAAS-CNRS, 7 avenue du colonel Roche, 31400 Toulouse (France)
simon.lacroix@laas.fr

Summary. This chapter presents an approach to control small sized airships to reach a given goal position, following a planned trajectory. Two sections are devoted to modeling and identification: a simplification of the airship dynamics when the airship speed is stabilized yields to a decoupling of the lateral and longitudinal dynamics. Various controllers to stabilize the speed, altitude and heading are proposed and analyzed. An investigation on the generation of feasible trajectories is proposed, and a controller dedicated to follow the generated trajectories is introduced. Simulations results illustrate the developments, as well as experimental results obtained with two different airships.

6.1 Introduction

The ever on-going developments in a wide spectrum of technologies, ranging from actuator, sensors and computing devices to energy and materials will ensure lighter than air machines a promising future. There is undoubtedly a regain of interest in this domain, as shown by recent industrial developments on heavy loads transportation projects[1], High Altitude Long Endurance platforms and surveillance applications. As for small size unmanned radio-controlled models, which size is of the order of a few tens of cubic meters, their domain of operation is mainly advertising or aerial photography. They are easy to operate, they can safely fly at very low altitudes (down to a few meters), and they can hover a long time over a particular area, while being able to fly at a few tens of kilometers per hour, still consuming little energy. Their main enemy is the wind (see [19] for a detailed and convincing review of the pros and cons of small size airships with regards to helicopters and planes). Let's also note that some specific applications of unmanned blimps are more and more seriously considered throughout the world, from planetary exploration to military applications, as shown by numerous contributions in the AIAA Lighter Than Air conferences and European Airship Conventions for instance [2, 1].

The first mentions of the development of unmanned autonomous blimps can be found in the literature of the early 90's [8], but it is only recently that various

* The authors would like to thank Alain Gauthier for his contributions to this Chapter.

[1] Such as the ATG Skycats – up-to-date information on this project is available at http://www.worldskycat.com

A. Ollero and I. Maza (Eds.): Mult. Hetero. Unmanned Aerial Vehi., STAR 37, pp. 147–188, 2007.
springerlink.com

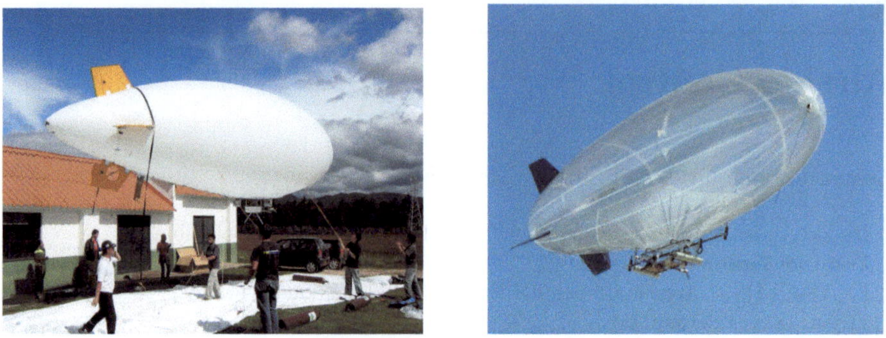

Fig. 6.1. The two airships UrAn (Andes University, $28\,m^3$) and Karma (LAAS, $18\,m^3$)

projects have reached effective achievements. One of the most advanced is Aurora, a project held at the Information Technology Institute of Campinas, Brazil, mainly devoted to flight control [15, 5]. Other projects are held at the University of Virginia [40], at the University of Stuttgart [32], the University of Wales [6] and the University of Evry [25]. An interesting characteristics of such projects is that they mix various innovative technological developments and fundamental research.

Chapter Overview and Outline

Flight control is of course the first problem to tackle in order to endow an airship with the ability to autonomously achieve any mission. Difficulties arise mainly because of the non-linearities of the airship model and the wind perturbations that strongly affects the flight. We focus here on this problem, aiming at defining means to control small sized airships to reach a given goal position, following a planned trajectory. Section 6.2 presents a complete dynamic airship model, and simplifications made to ease the synthesis of control laws. Under the hypotheses that the speed is regulated ("cruise flight hypotheses"), these simplifications allow to decouple the lateral and longitudinal dynamics, yielding two sub-models. Section 6.3 then presents means to estimate the parameters of the established models, so that proper control laws can be defined.

The overall approach to reach a goal while following a trajectory relies on three independent controllers that regulate the speed, heading and altitude, the heading reference being set by the path following control loop. Controls laws to stabilize speed, altitude and heading during cruise flight are presented in Sect. 6.4, and Sect. 6.5 presents how feasible trajectories can be planned and executed. References to previous work in the literature are presented throughout the sections, as well as simulation results and experimental results obtained either with the airships Karma and UrAn (Fig. 6.1).

6.2 Airship Modeling

The considered airships have a classic "cigar shaped" structure. The available control parameters are presented in Fig. 6.2: the main thrusters are mountedon

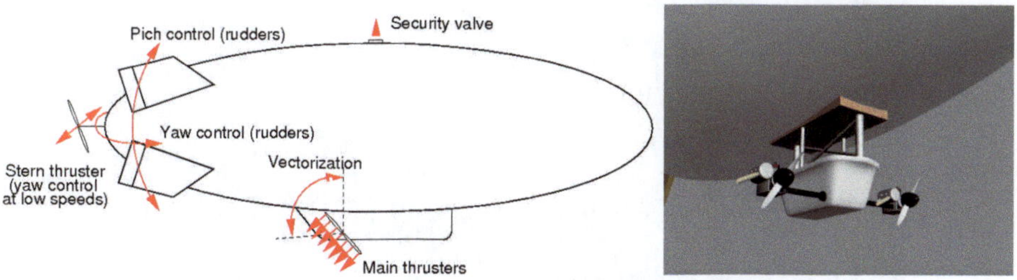

Fig. 6.2. The available control parameters. Right: detail of the vectorized thrusters of UrAn.

a vectorized axis, thus enabling vertical take off and providing additional lift at slow speeds, where no aerodynamic lift is possible. Rudders allows control in the longitudinal and lateral planes, and an additional tail rotor enables lateral control at slow speed, where the rudders control surfaces are not efficient[2].

6.2.1 Frames and Kinematic Model

Three frames are defined to describe the blimp motion (see Fig. 6.3). The global frame R_0 fixed to an arbitrary point on the earth, oriented along the NED convention. The airship body frame R_d has its origin at the center of volume (**CV**) of the hull [22]. **CV** is chosen as the origin of this frame because it is assumed to coincide with the center of buoyancy, where the aerostatic lift force is applied. Finally, the aerostatic frame aligned with the direction of motion is R_a.

The x_d axis of R_d is coincident with the symmetry axis of the envelope, the (x_d, z_d) plane coincides with the longitudinal plane and the orientation of R_d with respect to R_0 is given by the Euler angles roll ϕ, pitch θ and yaw ψ. The x_a axis of R_a is coincident with the airship aerodynamic velocity $v_{A|R_a} = (M_d^a)^{-1}(v_{dl} - v_w)$, where v_{dl} and v_w respectively denote the airship's velocity and the winds velocity with respect to R_d, and M_d^a is expressed as in (6.2). α is the angle of attack within the (x_d, y_d) plane and β is the skid angle within the (x_d, y_d) plane. The orientation matrix M_0^d between the global frame R_0 and local frame R_d is given by:

$$M_o^d = \begin{pmatrix} c\psi c\theta & -s\psi c\phi + s\phi c\psi s\theta & s\phi s\psi + s\theta c\psi c\phi \\ c\theta s\psi & c\psi c\phi + s\theta s\psi s\phi & -c\psi s\phi + c\phi s\theta s\psi \\ -s\theta & c\theta s\phi & c\theta c\phi \end{pmatrix} \tag{6.1}$$

and the transformation matrix M_d^a between the local frame R_d and the aeronautic frame R_a is:

$$M_d^a = \begin{pmatrix} c\alpha c\beta & -c\alpha s\beta & -s\alpha \\ s\beta & c\beta & 0 \\ s\alpha c\beta & -s\alpha s\beta & c\alpha \end{pmatrix} \tag{6.2}$$

where sx (resp. cx) denotes the function $\sin(x)$ (resp. $\cos(x)$).

[2] Only UrAn is actually equipped with such a thruster.

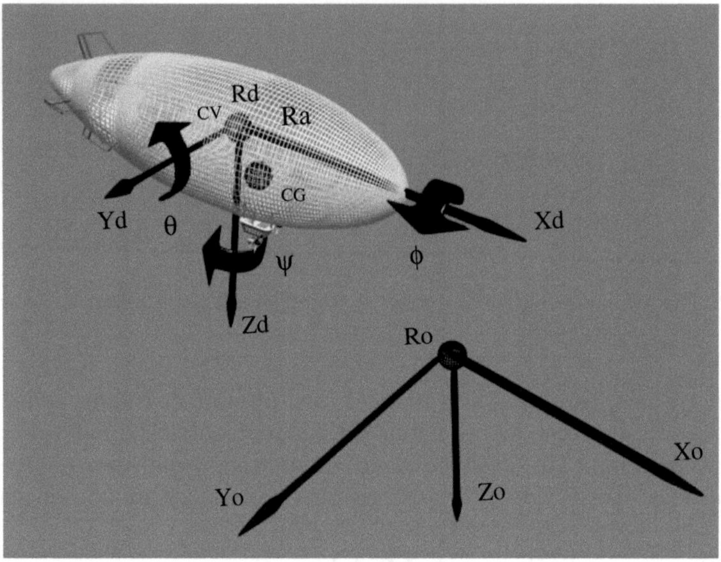

Fig. 6.3. Considered frames

6.2.2 Dynamic Model

The dynamic model of the airship is established on the basis of the following assumptions:

- The hull is considered as a solid: aero-elastic phenomena are ignored, and so are the motion of helium inside the hull (no phenomenon of inertial added fluid due to such motion is considered);
- the mass of the blimp and its volume are considered as constant;
- the blimp displaces volume: the phenomenon of added fluid induces a variation of inertia and mass and is significant (proportional to the volume of air displaced by the hull);
- the center of buoyancy is assumed to coincide with the CV;
- as the speed for a small blimp is generally low (low Match number), the couplings between dynamics and thermal phenomena are neglected and the density of air is not locally modified by the system's motion;
- the Earth is considered as flat over the flight area.

These assumptions are reasonable for the considered airships, and allow to apply the rigid body mechanics theory. Through the use of Newton's laws of mechanics, aero-dynamical theory [7] with Kirshoff's law and Bryson's theory, the dynamic model with respect to the R_d frame can be written as:

$$M_d\dot{\mathbf{v}} = -T_d(v_{dlr}) + T_a(v_A) + g_a + T_p \tag{6.3}$$

where:

- $\mathbf{v}$ is the airship's speeds states vector within R_d, which is composed of the linear velocities $v_{dl} = [u, v, w]^T$ and angular velocities $v_{dr} = [p, q, r]^T$. For control purpose, it is expressed in the frame R_0.
- the 6×6 matrix M_d is composed of masses, inertias and the corresponding coupling terms:

$$M_d = \begin{pmatrix} mI_3 & -m\tilde{A} \\ m\tilde{A} & I_N \end{pmatrix} = \begin{pmatrix} m & 0 & 0 & 0 & mz_g & 0 \\ 0 & m & 0 & -mz_g & 0 & mx_g \\ 0 & 0 & m & 0 & -mx_g & 0 \\ 0 & -mz_g & 0 & I_x & 0 & -I_{xz} \\ mz_g & 0 & -mx_g & 0 & I_y & 0 \\ 0 & mx_g & 0 & -I_{xz} & 0 & I_z \end{pmatrix} \tag{6.4}$$

where (x_g, z_g) is the gravity center position.
- T_d is the dynamic force vector and contains centrifugal and Coriolis terms:

$$T_d(v_{dlr}) = \begin{pmatrix} mwq - mvr - mx_g q^2 + mz_g pr - mx_g r^2 \\ mur - mwp + mz_g qr + mx_g pq \\ mvp - muq - mz_g p^2 + mx_g pr - mz_g q^2 \\ -mz_g ur + mz_g wp - I_{xz} pq - (I_y - I_z) qr \\ m(z_g wq - z_g vr - x_g vp + x_g uq) + I_{xz}(p^2 - r^2) - (I_z - I_x) pr \\ -mx_g wp + mx_g ur - (I_x - I_y) pq + I_{xz} qr \end{pmatrix} \tag{6.5}$$

- g_a gathers the gravity and buoyancy forces (F_G is the airship's weight, F_B is the buoyancy lift):

$$g_a = \begin{pmatrix} -(F_G - F_B) s\theta \\ (F_G - F_B) c\theta s\phi \\ (F_G - F_B) c\theta c\phi \\ -z_g F_G c\theta s\phi \\ -z_g F_G s\theta - (x_g F_G - x_c F_B) c\theta c\phi \\ (x_g F_G - x_c F_B) c\theta s\phi \end{pmatrix} \tag{6.6}$$

- T_p represents the controls applied on the airship: it contains the torques of the vectored thrust F_M at (O_x, O_z) position and of the tail rotor thrust F_{rc} at (x_{rc}, z_{rc}) position. The norm and the direction μ of the propellers are adjustable within the longitudinal plane and the direction of the tail rotor thrust within the lateral plane.

$$T_p = \begin{pmatrix} F_M c\mu \\ F_{rc} \\ -F_M s\mu \\ F_{rc} z_{rc} \\ F_M O_z c\mu + F_M O_x s\mu \\ F_{rc} x_{rc} \end{pmatrix} \tag{6.7}$$

- $T_a = A\dot{v_A} - D(v_{dr})v_A + T_{sta}(v_a^2)$ is the aerodynamic and moments vector, where:
  - A is the 6×6 symmetric matrix of added masses, inertia at center of gravity (**CG**), and coupling terms of the fluid (virtual mass and inertia terms – [39] presents how these coefficients can be estimated):

$$A\dot{v_A} = \begin{pmatrix} a_{11} & 0 & 0 & 0 & a_{15} & 0 \\ 0 & a_{22} & 0 & a_{24} & 0 & a_{26} \\ 0 & 0 & a_{33} & 0 & a_{35} & 0 \\ 0 & a_{42} & 0 & a_{44} & 0 & a_{46} \\ a_{51} & 0 & a_{53} & 0 & a_{55} & 0 \\ 0 & a_{62} & 0 & a_{64} & 0 & a_{66} \end{pmatrix} \begin{pmatrix} \dot{V}_{X_a|Rd} \\ \dot{V}_{Y_a|Rd} \\ \dot{V}_{Z_a|Rd} \\ \dot{p} \\ \dot{q} \\ \dot{r} \end{pmatrix} \tag{6.8}$$

  - $v_A = [v_a, v_{dr}]$, where $\mathbf{v}_a = \mathbf{v}_{dl} - \mathbf{v}_w$ is the vector of aerodynamic translational velocity, $\mathbf{v}_w = [u_w, v_w, w_w]^T$ being the wind velocity in R_d. A study on the influence of wind over airship can be found in [3].
  - $D(v_{dr})v_A$ is a vector that contains the centrifugal and Coriolis terms:

$$D_1(v_{dr})v_A = \begin{pmatrix} 0 & a_{22}r & -a_{33}q & a_{24}r & -a_{35}q & a_{26}r \\ D_{1z1} & 0 & a_{33}p & a_{35}q & -a_{15}q & 0 \\ (a_{11} - xm_{22})q & -a_{22}p & 0 & -a_{24}p - a_{26}r & a_{15}q & 0 \\ D_{1z2} & -(a_{62} + a_{35})q & D_{1z3} & -a_{64}q & (a_{55} - a_{66})r & 0 \\ (a_{35} + x^2m_{22})q & -a_{42}ra_{62}p & -a_{15}q & D_{1z5} & 0 & -a_{64}r \\ D_{1z4} & (a_{15} + a_{42})q & -a_{53}p & (a_{44} - a_{55})q & a_{46}r & 0 \end{pmatrix} \begin{pmatrix} V_{X_a|Rd} \\ V_{Y_a|Rd} \\ V_{Z_a|Rd} \\ p \\ q \\ r \end{pmatrix} \tag{6.9}$$

  where $D_{1z1} = pm_{13} + r(xm_{11} - a_{11})$, $D_{1z2} = pm_{33} + r(a_{15} + xm_{13})$, $D_{1z3} = (a_{62} + a_{35})r + a_{24}p$, $D_{1z4} = -(a_{51} + a_{24} - xm_{13})p - (a_{26} - x^2m_{11})r$ y $D_{1z5} = a_{64}p + (a_{66} - a_{44})r$.
  - $T_{sta}(v_a^2)$ represents the aerodynamic forces and moments at CG, which are proportional to the form of the hull and the square of the aerodynamic velocity.

$$T_{sta}({v_a}^2) = \begin{pmatrix} \frac{1}{2}\rho {v_a}^2 S_{ref} C_T \\ \frac{1}{2}\rho {v_a}^2 S_{ref} C_L \\ \frac{1}{2}\rho {v_a}^2 S_{ref} C_N \\ -\frac{1}{2}\rho {v_a}^2 S_{ref} L_{ref} C_l \\ -\frac{1}{2}\rho {v_a}^2 S_{ref} L_{ref} C_m \\ -\frac{1}{2}\rho {v_a}^2 S_{ref} L_{ref} C_n \end{pmatrix} = \begin{pmatrix} F_x \\ F_y \\ F_z \\ L_0 \\ M_0 \\ N_0 \end{pmatrix} \tag{6.10}$$

where S_{ref} and L_{ref} depend of the airship geometry. C_T, C_L, C_N, C_l, C_m and C_n are, respectively, tangential, normal, lateral, roll, pitch and yaw stationary coefficients.

6.2.3 Simplified Model

In order to obtain a tractable dynamic model, additional simplifications can be made, by restricting the domain of operation and decoupling lateral and longitudinal controls. In the case of a cruise flight in the absence of wind, the equation that links the blimp velocity and thrust control comes down to:

$$\dot{u} = \frac{1}{(m - a_{11})}(F_m \cos\mu + \frac{1}{2}\rho V_a^2 S_{ref} C_T) \tag{6.11}$$

Provided that the airship speed is regulated ($u = u_0$) and that it is flying along a straight line ($\delta_g = 0$, $v = 0$), the evolution model of the altitude can be simplified. The involved variables in the model are z, w, u, θ, q and δ_e: considering that $w \ll u$ and $\alpha \approx 0$, we have:

$$\begin{gathered} \dot{z} = -u \sin\theta \\ \dot{\theta} = q \\ \dot{q} = \tfrac{1}{I_y - a_{55}}[(mx_g - a_{35} - x^2 m_{22})uq - z_g F_G \sin\theta - (x_g F_G - x_c F_B)\cos\theta + \\ + F_m O_z \cos\mu + F_m O_x \sin\mu - \tfrac{1}{2}\rho u^2 S_{ref} L_{ref} C_{mN}] \end{gathered} \tag{6.12}$$

By linearizing around the operating point, $-(x_g F_G - x_c F_B)\cos\theta + F_m O_z \cos\mu \approx 0$, $\theta \approx 0 \rightarrow \sin\theta \approx \theta$ and $\cos\theta \approx 1$, we have:

$$\begin{gathered} \dot{z} = -u \sin\theta \\ \dot{\theta} = q \\ \dot{q} = k_{2|\delta_e} q + k_{1|\delta_e}\theta + k_{3|\delta_e}\delta_e \end{gathered} \tag{6.13}$$

where $k_{2|\delta_e} = \frac{(mx_g - a_{35} - x^2 m_{22})u}{I_y - a_{55}}$, $k_{1|\delta_e} \approx \frac{-z_g F_G}{I_y - a_{55}}$ and $k_{3|\delta_e} \approx -\frac{\rho u^2 S_{ref} L_{ref} C_{mN}}{2(I_y - a_{55})}$. Equation (6.13) corresponds to a second order control system in θ, and a third order system in z. Considering the dynamic of θ similar as a first order, the model of the altitude z can be seen as a first order system plus an integral term.

Considering now a constant speed motion in the plane (x, y) with $\delta_e = 0$ and $w = 0$, the dynamic equations of the blimp and its position in the horizontal plane can be written as:

$$\begin{gathered} \dot{x} = u \cos\psi - v \sin\psi \\ \dot{y} = u \sin\psi - v \cos\psi \\ \dot{\psi} = r \\ \dot{r} = \frac{ur(mx_g + a_{26} - x^2 m_{11}) + F_{rc} x_{rc} - \frac{1}{2}\rho u^2 S_{ref} L_{ref} C_{nN}}{(I_z - a_{66})} \end{gathered} \tag{6.14}$$

The involved variables in this model are x, y, u, v, ψ, r and δ_g. Considering $v \ll u$, $\beta \approx 0$ and $F_{rc} = 0$, the equations (6.14) can be simplified in:

$$\begin{gathered} \dot{x} = u \cos\psi \\ \dot{y} = u \sin\psi \\ \dot{\psi} = r \\ \dot{r} = k_{2|\delta_g} r + k_{1|\delta_g}\delta_g \end{gathered} \tag{6.15}$$

Note that the coefficients $k_{1|\delta_g}$ and $k_{2|\delta_g}$ can be found similarly to the above altitude model reduction procedure (a similar application of simplified dynamics for an airplane model in cruise flight can be found in [21]).

6.3 Model Identification

One way to estimate the aerodynamic model parameters is the use of wind-tunnels experiments with Munk's hull equations [35]. Jones and DeLaurier [30] estimate the aerodynamic coefficients of an airship from some theoretical results and wind-tunnel data. Gomes and Ramos [22] propose that the virtual masses and inertia can be basically estimated with the Munk's hull equations. Furthermore, they show that the aero-dynamic coefficients of equation (6.10) can be obtained either from direct measurement in a wind-tunnel, from the geometrical characteristics of the blimp, or from aerodynamic stability derivatives. The coefficients proposed by Hygounenc and Souères [28] have been determined by means of wind-tunnel tests made with a scaled model of Karma (see photo 6.4).

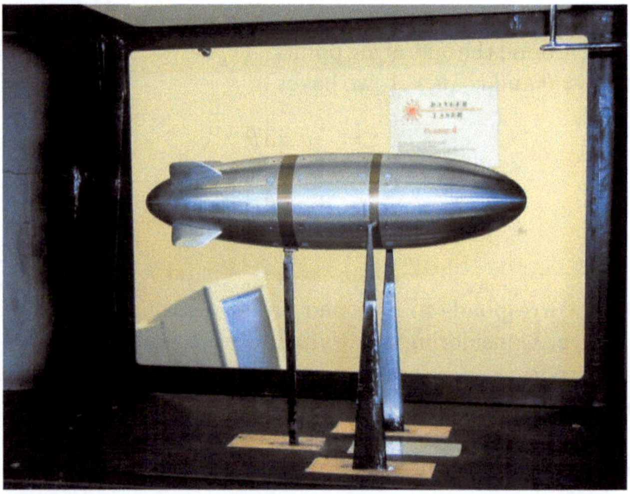

Fig. 6.4. Scale model of Karma

Few publications address the estimation of airship aerodynamical parameters from real world flight data. The dynamic identification of an airship is presented in [29], where the authors develop a general aerodynamic model at the reference operation condition of the airship, based on Munk's theory and Jones and DeLaurier's contributions. Yamasaky and Goto [41] with the objective to construct models for the development of control laws, present two experiments in order to identify the airship flight dynamics: these experiments consist in constrained flight tests and indoors free-flying tests.

In the following section (Sect. 6.3.1), we present an approach to estimate the aerodynamic parameters from cruise flight data. Section 6.3.2 analyzes the parametric variations and the validity of the models simplifications presented in Sect. 6.2.

6.3.1 Estimation of the Aerodynamic Parameters

To avoid experimental aerodynamic measurements in wind tunnels, we adopted a numerical technique that determines the system coefficients (λ) by fitting time series data obtained from state measurements of the system. The difficulty arises from the fact that the acquired data are corrupted by noise: we present here an approach to parameters identification based on the reduction of $T_a(v_A)$ when the airship is in cruise flight, using the Kalman filter as a parameters identifier. The unknown parameters are in the state vector of the filter, and are constant in the particular case of a steady state regime: $\lambda_k = \lambda_{k-1}$. Therefore, the state space is:

$$\begin{aligned} x_{k+1} &= f(x_k, \lambda_k, u_k) + \epsilon_k \\ \lambda_k &= \lambda_{k-1} + \epsilon_i \end{aligned} \tag{6.16}$$

where x, y, u and λ are, respectively, the state system, the output system, the input system and the unknown coefficients, and $\epsilon_{k,i}$ are gaussian noises. The observation function is:

$$y_k = h(x_k) + \nu_k \tag{6.17}$$

The expression of $T_a(v_A)$ is cumbersome. With the assumptions presented in Sect. 6.2 (cruise flight), we can simplify $T_a(v_A)$ and still obtain a general representation which includes the aerodynamic phenomena:

$$T_a(v_A) = \begin{pmatrix} k1(c_{T1} + c_{T2}\alpha + c_{T3}\delta_e) \\ k1(c_{L1} + c_{L2}\beta + c_{L3}\delta_g) \\ k1(c_{N1} + c_{N2}\alpha + c_{N3}\beta) \\ k2(c_{l1} + c_{l2}\alpha + c_{l3}\delta_e) \\ k2(c_{m1} + c_{m2}\beta + c_{m3}\delta_g) \\ k2(c_{n1} + c_{n2}\alpha + c_{n3}\beta) \end{pmatrix} \tag{6.18}$$

where $k_1 = \frac{1}{2}\rho S_{ref}$ and $k_2 = \frac{1}{2}\rho S_{ref} L_{ref}$ are constants depending on the airship geometry. This simpler model of the aerodynamics of the airship in cruise flight also simplifies the formulation of the control laws.

Estimation in Simulation

To determine the performance and reliability of our method, we have carried out simulation tests, comparing the results of the Extended Kalman Filter (EKF) and the Unscented Kalman Filter (UKF, in [14] the authors present an application of UKF on spacecraft attitude estimation) as parametric identifiers, with known system parameters. These parameters, that we are going to take as real, were identified by wind-tunnel tests, as Hygounenc and Souères present in [28].

Table 6.1. Table of error

State	Mean UKF	Variance UKF	Mean EKF	Variance EKF
u	0.0177	0.0004	0.0190	0.0006
v	0.0034	0.0006	0.0058	0.0007
w	0.0156	0.0003	0.0287	0.0007
p	0.0016	0.0001	0.0017	0.0002
q	0.0019	0.0001	0.0023	0.0002
r	0.0057	0.0001	0.0020	0.0002
x	0.0439	0.0015	0.0639	0.0016
y	0.0670	0.0025	0.0209	0.0027
z	0.0217	0.0038	0.0086	0.0040
ϕ	0.0626	0.0009	0.0651	0.0009
θ	0.0179	0.0001	0.0209	0.0003
ψ	0.0153	0.0171	0.0603	0.0018

We use $N = 1500$ samples and the Euler method to integrate the differential equations with a step size $\Delta T = 0.1$. Notice that the integration step size may differ from the sampling time. The state measurements are corrupted with an additive Gaussian noise (similar to the on-board real sensors). Results are shown in the left side of Table 6.2, and Table 6.1 shows the mean and variance of the error $e = \frac{|x-\hat{x}|}{|x|}$ for the blimp state (speed vector $[u, v, w, p, q, r]^T$ in R_d and position vector $[x, y, z, \phi, \theta, \psi]^T$ in R_0). For both model and parameters, the UKF approach converges more rapidly to a better estimate than the EKF.

Estimation with Real World Data for Karma

Once made the study of the algorithms with simulation data, we use real data acquired by the GPS and the compass on-board Karma, that respectively provides estimates of $(x, y, z, \dot{x}, \dot{y}, \dot{z})$ and (ϕ, θ, ψ) expressed in R_0, at a GPS rate of 1 Hz and a compass rate of 10 Hz. Data acquired during a 923 seconds long trajectory are exploited to estimate the airship model parameters using an UKF. The right side of Table 6.2 shows all the estimated values of the model parameters of the Karma with their covariances (P_{ii}).

6.3.2 Analysis of the Validity of the Reduced Models

The approach to airship parameters estimation in cruise flight, presented in Sect. 6.3.1, enables to know the parameters of simplified models. To make sure the control laws derived from the reduced models are sound, an analysis of the reduced models validity is required. The classical techniques used for system identification such step response and extended least squares enable to say that the airship global dynamics can be represented by the simplified models. This section presents such an analysis, using an Extended Kalman Filter to estimate the state of each simplified model and the evolution of their parameters.

Table 6.2. Left side: Parameter estimation from simulated data and right side: Parameter estimation from real world data

Param.	Real	EKF	UKF
a_{11}	1.5276	1.2633	1.9660
a_{22}	21.0933	20.8660	21.1896
a_{33}	20.4220	20.2018	21.4239
a_{44}	16.3905	15.9151	16.8548
a_{55}	382.1290	380.0030	380.1463
a_{66}	388.0972	379.9988	384.0155
$a_{15} = a_{51}$	0	0.1309	0.0001
$a_{24} = a_{42}$	0	−0.0958	0.1628
$a_{26} = a_{62}$	−69.6939	−59.9030	−69.8978
$a_{35} = a_{53}$	67.7045	70.0684	70.3361
a_{46}	0	0.0801	0.0101
m_{13}	1.2801	2.0621	1.5235
m_{33}	−49.7019	−48.0249	−48.5013
xm_{11}	25.6919	23.0748	24.5183
xm_{22}	23.6878	20.0129	21.0075
xm_{13}	−4.5582	−9.1165	−5.4699
x^2m_{11}	−173.4906	−150.0044	−170.8227
x^2m_{22}	−166.3538	−149.9994	−158.8524
C_{T1}	–	−1.8974	−0.6579
C_{T2}	–	0.1071	0.1069
C_{T3}	–	−2.8752	−0.6877
C_{L1}	–	50.8756	15.4789
C_{L2}	–	−15.9874	−11.5582
C_{L3}	–	−1.2234	−7.2243
C_{N1}	–	−0.0417	−0.0664
C_{N2}	–	0.5487	0.5789
C_{N3}	–	1.0258	0.389
C_{l1}	–	0.0205	0.0304
C_{l2}	–	−0.1919	−0.1271
C_{l3}	–	0.0170	0.0266
C_{m1}	–	−0.0405	−0.0415
C_{m2}	–	0.7975	0.9982
C_{m3}	–	0.2930	0.2173
C_{n1}	–	1.0833	0.6207
C_{n2}	–	−0.8574	−0.7589
C_{n3}	–	0.0450	−0.0010

Param.	UKF	Cov.(P_{ii})
a_{11}	11.2428	0.1580
a_{22}	90.4922	0.0925
a_{33}	70.5841	0.0922
a_{44}	42.3387	0.0901
a_{55}	383.7979	0.0837
a_{66}	419.9314	0.0872
$a_{15} = a_{51}$	6.9103	0.1309
$a_{24} = a_{42}$	1.2382	0.1240
$a_{26} = a_{62}$	−195.3407	0.1269
$a_{35} = a_{53}$	59.4323	0.1053
a_{46}	−28.5030	0.1053
m_{13}	33.7772	0.0621
m_{33}	−93.7707	0.0982
xm_{11}	76.4093	0.0905
xm_{22}	54.7163	0.0192
xm_{13}	75.3240	0.0962
x^2m_{11}	−201.9972	0.0335
x^2m_{22}	−224.8353	0.0896
C_{T1}	−2.9074	0.1290
C_{T2}	−0.2250	0.0446
C_{T3}	−0.7970	0.0767
C_{L1}	15.0799	0.0744
C_{L2}	−7.6177	0.0644
C_{L3}	−3.2706	0.0249
C_{N1}	−2.1196	0.0676
C_{N2}	−13.9818	0.0949
C_{N3}	0.6837	0.0508
C_{l1}	5.1576	0.0538
C_{l2}	2.9208	0.1509
C_{l3}	1.0168	0.0582
C_{m1}	−0.0725	0.1442
C_{m2}	−1.8937	0.0814
C_{m3}	1.1017	0.0762
C_{n1}	−0.1082	0.0942
C_{n2}	−0.5101	0.0415
C_{n3}	0.0115	0.0227

Speed vs. Thrust Propeller v_a/T

Let us consider the model (6.11) with the state estimation $\chi = [x, \kappa, \tau]^T$. The corresponding Euler model is given by:

$$
\begin{aligned}
x(k+1) &= x(k) + \tfrac{T_m}{\tau(k)}(\kappa(k)u(k) - x(k)) \\
\tau(k+1) &= \tau(k) \\
\kappa(k+1) &= \kappa(k)
\end{aligned}
\qquad (6.19)
$$

It is possible to demonstrate that the matrix F (the linearized form of f of the state equation) can be written as:

$$F = \begin{pmatrix} (1 - \frac{T_m}{\tau}) & \frac{T_m u}{\tau} & \frac{T_m}{\tau^2}(-\kappa u + x) \\ 0 & 1 & 0 \\ 0 & 0 & 1 \end{pmatrix} \tag{6.20}$$

Figure 6.5 shows the evolution of the state estimation and Fig. 6.6 presents a stability and statistic analysis obtained in simulation.

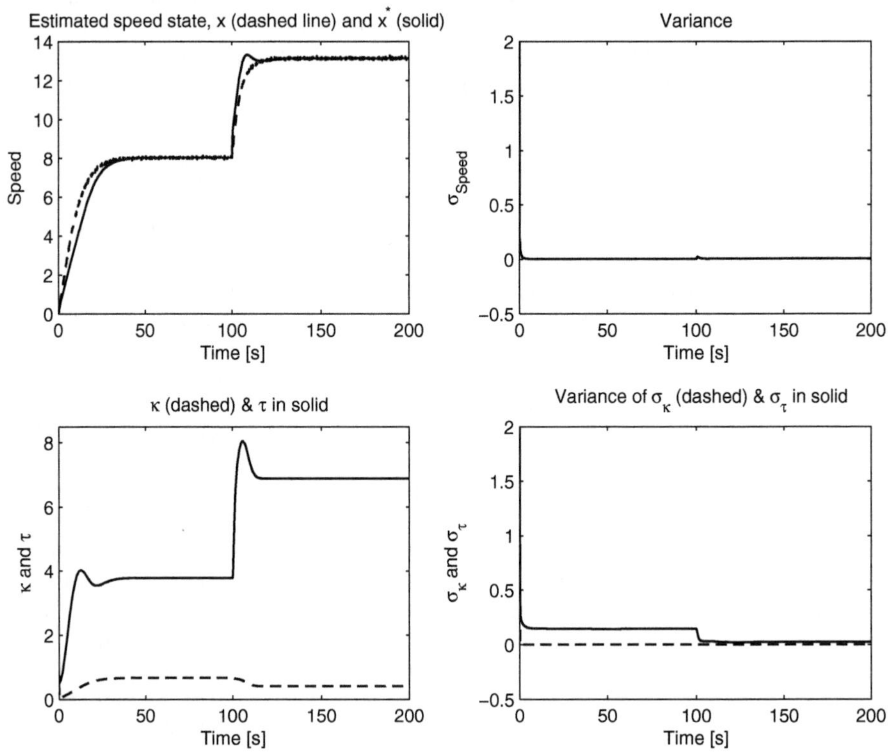

Fig. 6.5. Evolution of speed, τ and κ states

Altitude vs. Symmetrical Deflection z/δ_e

Considering the model (6.13) with $\chi_1 = [z, \theta, q, k_z, k_1, k_2, k_3]^T$ as states for the estimation, the Euler model can be written as:

$$\begin{aligned} z(k+1) &= z(k) + T_m k_z(k)\theta(k) \\ \theta(k+1) &= \theta(k) + T_m q(k) \\ q(k+1) &= q(k)(1 + T_m k1(k)) + T_m(k_2(k)\theta(k) + k_3(k)\delta_e(k)) \end{aligned} \tag{6.21}$$

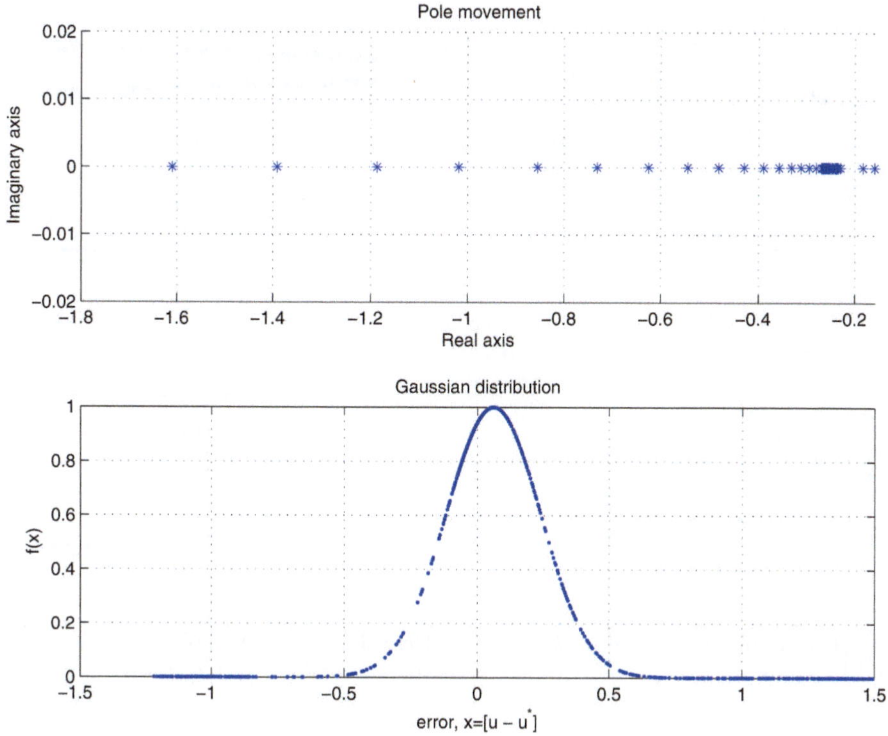

Fig. 6.6. Analysis: poles movement and statistic error of speed state

The F matrix is given by:

$$F = \begin{pmatrix} 1 & T_m k_z & 0 & T_m\theta & 0 & 0 & 0 \\ 0 & 1 & T_m & 0 & 0 & 0 & 0 \\ 0 & T_m k_2 & (1+T_m k1) & 0 & T_m q & T_m\theta & T_m\delta_e \\ 0 & 0 & 0 & 1 & 0 & 0 & 0 \\ 0 & 0 & 0 & 0 & 1 & 0 & 0 \\ 0 & 0 & 0 & 0 & 0 & 1 & 0 \\ 0 & 0 & 0 & 0 & 0 & 0 & 1 \end{pmatrix} \tag{6.22}$$

Figures 6.7 and 6.8 present the results obtained. They show the parametric convergence of the model.

Heading vs. Dis-symmetrical Deflection ψ/δ_g

For the model (6.15), the states for the estimation $\chi_1 = [\psi, r, k_1, k_2]^T$, and the Euler model is given by:

$$\begin{aligned} \psi(k+1) &= \psi(k) + T_m r(k) \\ r(k+1) &= r(k)(1 + T_m k_2(k)) + T_m k_1(k)\delta_g(k) \end{aligned} \tag{6.23}$$

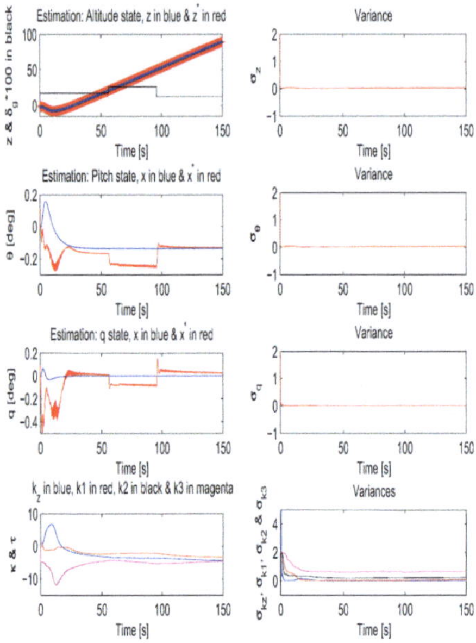

Fig. 6.7. Evolution of z, θ, q, k_z, k_1, k_2 and k_3 states

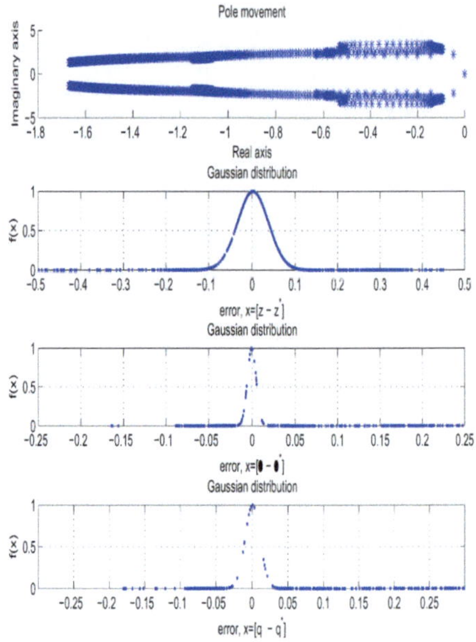

Fig. 6.8. Analysis: poles movement and statistic error of z, θ, q states

The F matrix can be written as:

$$F = \begin{pmatrix} 1 & T_m & 0 & 0 \\ 0 & (1 + T_m k_2) & T_m \delta_g & T_m r \\ 0 & 0 & 1 & 0 \\ 0 & 0 & 0 & 1 \end{pmatrix} \tag{6.24}$$

The system states (ψ and r) are presented in Fig. 6.9, and an analysis of the estimation errors and pole movements can be seen in Fig. 6.10.

In summary, Figs. 6.5, 6.7 and 6.9 respectively show the validity of the simplified speed, altitude and heading models: the estimated parameters converge to a fixed values. Figures 6.6, 6.8 and 6.10 show that the poles remain in the left side of the complex plane when the system evolves in cruise flight.

Once the simplified model study had been done and the test of the algorithms on simulation had been made, we used real flight data. Figures 6.11 and 6.12 show the speed state evolution and the corresponding stability analysis when the airship was moving in the cruise flight phase.

6.4 Control

Now that we have a tractable model with estimates of its coefficients, one can tackle the control problem. Airship flight control issues have now been addressed in various contributions in the literature. Elfes et al. [20] present a control architecture where the yaw, pitch and altitude of the airship are controlled on the basis of adaptive adjustments of PI and PID controllers. Paiva et al. [16] describe a control strategy with a PID controller for the longitudinal velocity, and a PD controller for altitude and heading. Azinheira et al. [4] present a heading regulator based on H_∞ control technique, exploited in a path following strategy. Moutinho and Azinheria [34] present a dynamic inversion controller (feedback linearization) for the lateral and longitudinal control, and an other feedback linearization approach combined with PID controllers has been proposed by Zhang and Ostrowski [43]. Some authors opt for fuzzy control approaches, such as Rao et al. [23], who describe a heading controller for navigation task based on fuzzy control optimized by genetic algorithms, and Diaz et al. [17] who propose a parallel distributed compensation for cruise flight phase based on fuzzy techniques. A robust stability augmentation system for yawing motion has been presented by Yong-Hwan Mo et al. [42]. A control strategy with back-stepping techniques was proposed by Hygounenc and Souères [28].

In this section, we present and analyze three different types of controls for the airship: PID control, Generalized Predictive Control (GPC) and non-linear control by extended linearization. All these regulators are conceived as Single Input-Single Output (SISO) systems. These controllers are established on a decomposition of the system into two independent lateral and longitudinal subsystems [26], and are the following:

- a control of the aerodynamic speed acting on the main thrusters,
- an altitude controller acting on the tail-fins by the δ_e signal,
- and a heading controller acting on the tail-fins by the δ_g signal.

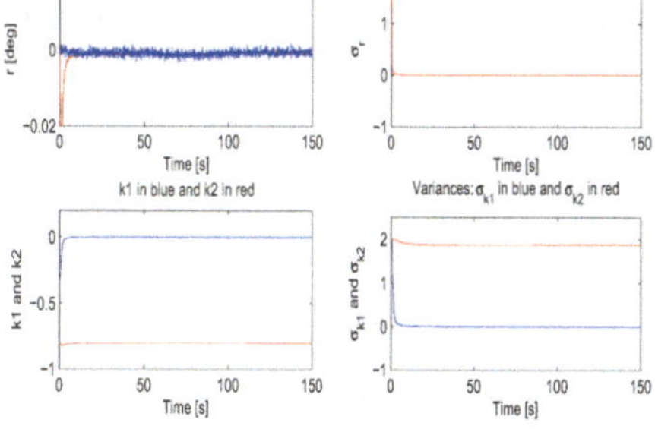

Fig. 6.9. Evolution of ψ, r, k_1 and k_2 states

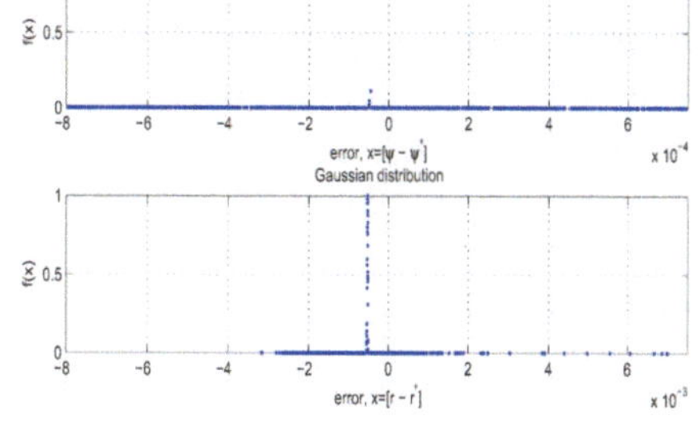

Fig. 6.10. Analysis: poles movement and statistic error of ψ and r states

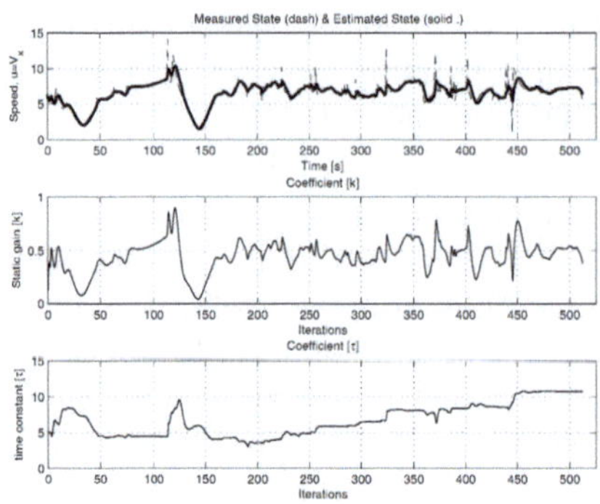

Fig. 6.11. Evolution of speed, τ and κ states obtained on real world data

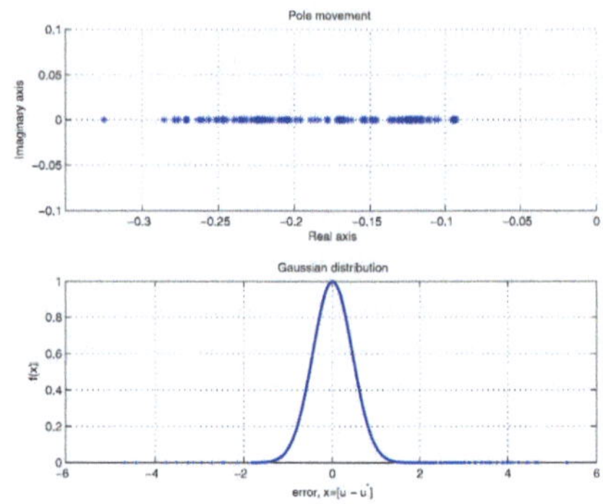

Fig. 6.12. Poles movement and statistic error of speed state based on real world data

Thus, the problem consists in designing control laws so that the airship output will follow as closely as possible the reference signals u_r, z_r and ψ_r, respecting the cruise flight hypotheses for the validity of simplified models, that is a similar behavior to the answer in open loop (respect to the natural movement of the airship).

6.4.1 PID Control

Every simplified model presented in Sect. 6.3.2 can be represented by:

$$\begin{aligned} \dot{x} &= Ax + Bu \\ y &= Cx \end{aligned} \tag{6.25}$$

where x is a $n \times 1$ state vector, y is a $p \times 1$ output vector and u is a $q \times 1$ input vector (A, B, C and D are respectively, $n \times n$, $n \times p$, $q \times n$ and $q \times p$ matrices). The control law u defined by a PID controller can be written as:

$$u = K_p u_m + K_i \int (u_{ref} - u_m)\mathrm{d}t + K_d \frac{\mathrm{d}u_m}{\mathrm{d}t} \tag{6.26}$$

where u_{ref} is the reference and u_m is the controlled variable. The proportional, integral and derivative gains K_p, K_i and K_d can be determined thanks to state feedback and pole placement theory.

Finally, it is recommended to use an anti-windup strategy to avoid saturation of the integral action term. The scheme that represents the PID control with its saturation is presented in Fig. 6.13

Speed Controller

With the open-loop control model for the aerodynamic speed written as

$$H(s) = \frac{v_a(s)}{T(s)} = \frac{\kappa}{\tau s + 1} \tag{6.27}$$

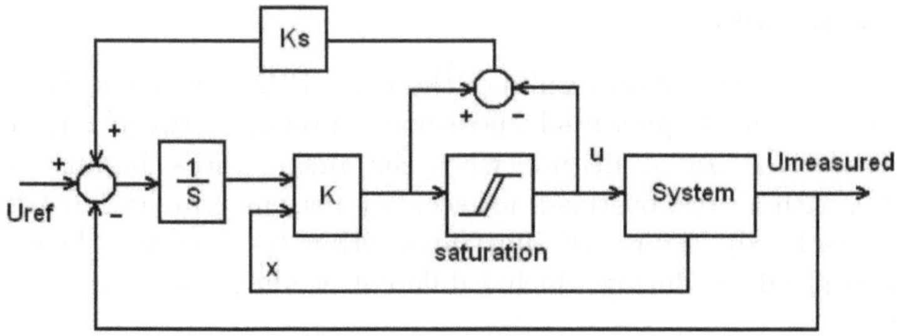

Fig. 6.13. PID regulator scheme

and the control input T (thrust propeller) given by

$$T(s) = k_p v_a + \frac{k_i}{s}(v_r - v_a) \tag{6.28}$$

where v_r is the reference speed, the closed-loop characteristic polynomial becomes

$$s^2 - \frac{1}{\tau}(1 + \kappa k_p)s + \frac{\kappa}{\tau}k_i \tag{6.29}$$

Now, defining the polynomial $s^2 + 2\zeta w_n s + w_n^2$ based on performance criteria and equaling with (6.29), the coefficients k_p and k_i can be found.

Altitude and Heading Controller

With a control model for the altitude written as

$$H(s) = \frac{z(s)}{\delta_e(s)} = \frac{k_z k_3}{s^3 - k_2 s^2 - k_1 s} \tag{6.30}$$

and the control input

$$\delta_e(s) = k_p z + \frac{k_i}{s}(z_r - z) + k_d\theta + k_{2d}q \tag{6.31}$$

where z_r is the reference altitude, the closed-loop characteristic polynomial becomes

$$s^4 - s^3(k_2 + k_3 k_{2d}) - s^2(k_1 + k_3 k_d) - s(k_p k_3 k_z) + k_z k_3 k_i \tag{6.32}$$

Based on the performance criteria a polynomial $(s^2+2\zeta w_n s+w_n^2)(s+1/\tau_1)(s+1/\tau_2)$ can be formulated. Equaling it with (6.32), the coefficients k_p, k_d, k_i and k_{2d} to control the altitude can be determined.

The gains to control the heading are derived with the same formulation.

Simulation Results

Figure 6.14 shows simulation results of the three PID controllers. Stabilization of each output state is presented and some couplings between regulators are showed. There is a lost in altitude when the airship turns, because it always made flights with a little overload, for security reasons (gravity force is greater than buoyancy force). Figure 6.15 illustrates within 3D space how the controllers stabilize a cruise flight during which 4 different heading references are given by an operator.

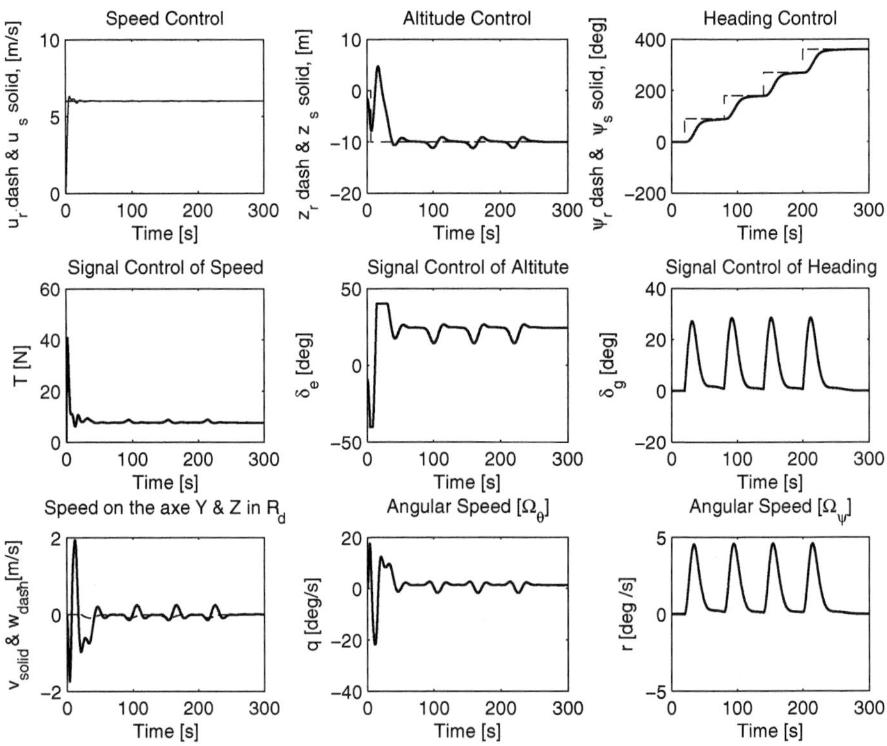

Fig. 6.14. Results of the PID controllers: from top to bottom, evolution of the controlled parameter with respect to the reference, evolution of the control input, and evolution of the derivatives of the controlled parameters

6.4.2 Generalized Predictive Control

General Predictive Control was introduced by Clarke et al. [13], and has become popular in both industry and academy [12, 11, 33]. It is a class of predictive control based on the minimization of an objective function to generate the control law.

A SISO system can be described by the following general form:

$$A(z^{-1})y(t) = z_{-d}B(z^{-1})u(t-1) + C(z^{-1})e(t) \tag{6.33}$$

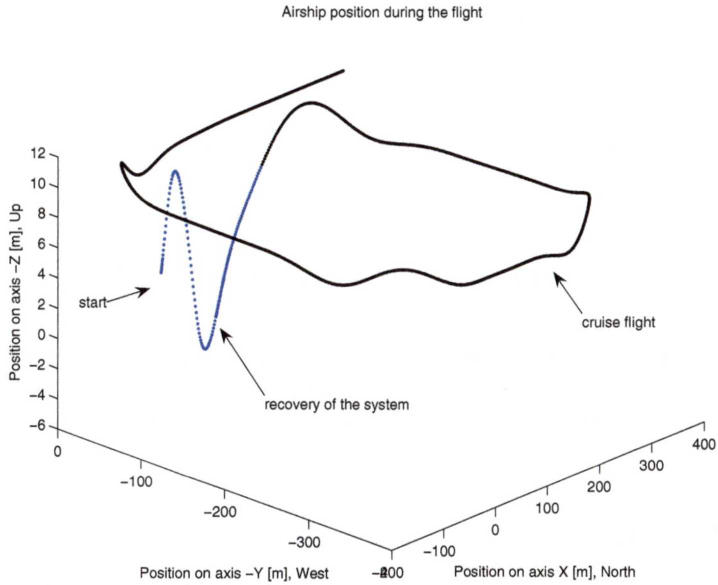

Fig. 6.15. Motion resulting from the application of the three PID controllers, with the states stabilized in Fig 6.14

where $u(t)$ and $y(t)$ are the control and output signals of the system and $e(t)$ is a zero mean white noise. A, B and C are polynomials with the z operator and d is the dead time of the system. This model is called "Controller Auto-Regressive Moving-Average (CARMA)" model. When the disturbances are non-stationary, the CARMA model is not appropriated and we must consider an integrated CARMA model (CARIMA) [13]. Considering the same notation followed by Camacho [11] the CARIMA model becomes:

$$A(z^{-1})y(t) = z_{-d}B(z^{-1})u(t-1) + C(z^{-1})\frac{e(t)}{\Delta} \tag{6.34}$$

with

$$\Delta = 1 - z^{-1} \tag{6.35}$$

The GPC algorithm consists of applying a control sequence that minimizes a cost function of the form:

$$J(N_1, N_2, N_u) = E\{\sum_{j=N_1}^{N_2} \delta(j)[y(t+j|t) - w(t+j)]^2 + \sum_{j=1}^{N_u} \lambda(j)[\Delta u(t+j-1)]^2\} \tag{6.36}$$

where $E\{*\}$ is the expectation operator and $\hat{y}(t+j|t)$ is an optimum j step ahead prediction of the system output on data up to time t, N_1 and N_2 are the minimum and maximum costing horizons, N_u is the control horizon, $\delta(j)$ and $\lambda(j)$ are weighting sequences and $w(t+j)$ is the future reference trajectory.

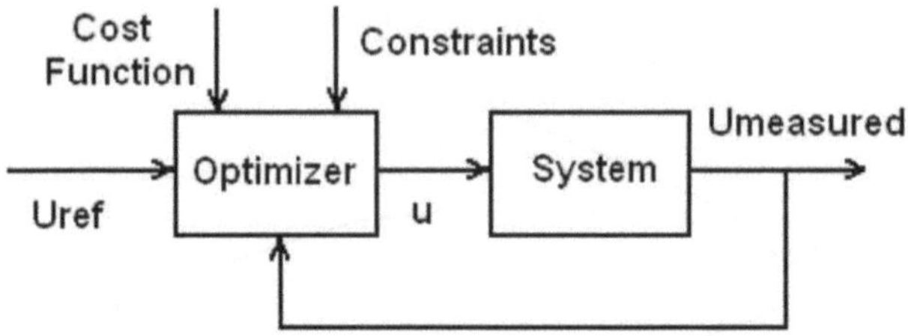

Fig. 6.16. GPC regulator scheme

The objective of GPC control is then to find the future control sequence $u(t)$, $u(t+1)$, ... so that the future system output $y(t+j)$ will follow as closely as possible $w(t+j)$, of course by minimizing the cost function $J(N_1, N_2, N_3)$ (see Fig. 6.16).

Speed Controller

The following discrete equivalence can be obtained when the simplified speed model is discretized with sample time $t_s = 0.1$ s:

$$H(z^{-1}) = \frac{0.0198z^{-1}}{1 - 0.9802z^{-1}} \tag{6.37}$$

Note that for this system the delay d is equal to zero and the noise polynomial is equal to 1.

Now with the algorithm presented by Camacho [11] and the parameter values $N_1 = 1$, $N_2 = N = 10$ and $\lambda = 1.2$ the control law becomes:

$$\begin{aligned} T(k) = T(k-1) - 4.7404v_a(k) + 4.0845v_a(k-1) + 0.0151v_{a|ref}(k+1) \\ +0.0288v_{a|ref}(k+2) + \cdots + 0.1084v_{a|ref}(k+10) \end{aligned} \tag{6.38}$$

Altitude Controller

Let us consider the following discrete altitude model:

$$H(z^{-1}) = \frac{0.000603 + 0.002424z^{-1} + 0.000609z^{-2}}{1 - 3.021z^{-1} + 3.041z^{-2} - 1.02z^{-3}} \tag{6.39}$$

If $d = 0$, $C(z^{-1}) = 1$ and the parameter values $N_1 = 1$, $N_2 = N = 10$ and $\lambda = 15$, the calculated control becomes:

$$\begin{aligned} \delta_e(k) = 0.9484\delta_e(k-1) - 0.0516\delta_e(k-2) - 24.2363z(k) + \\ +65.6447z(k-1) - 59.8995z(k-2) + 18.3751z(k-3) + \\ 0.00004z_{ref}(k+1) + 0.00031z_{ref}(k+2) + \cdots + 0.0384z_{ref}(k+10) \end{aligned} \tag{6.40}$$

Heading Controller

Considering the discrete heading model:

$$H(z^{-1}) = \frac{0.001839z^{-1} + 0.001722z^{-2}}{1 - 1.822z^{-1} + 0.8219z^{-2}} \tag{6.41}$$

and knowing $d = 0$, $C(z^{-1}) = 1$, the parameter values $N_1 = 1$, $N_2 = N = 10$ and $\lambda = 0.6$, the heading control law is written as:

$$\begin{aligned} \delta_g(k) = 0.9717\delta_g(k-1) - 0.0282\delta_g(k-2) - 38.5727\psi(k) + \\ +64.3467\psi(k-1) - 26.4822\psi(k-2) + 0.0029\psi_{ref}(k+1) + \\ +0.0108\psi_{ref}(k+2) + \cdots + 0.1609\psi_{ref}(k+10) \end{aligned} \tag{6.42}$$

Simulation Results

Figures 6.17 and 6.18 show the results of the GPC controller, with the same airship model and references as in Sect. 6.4.1.

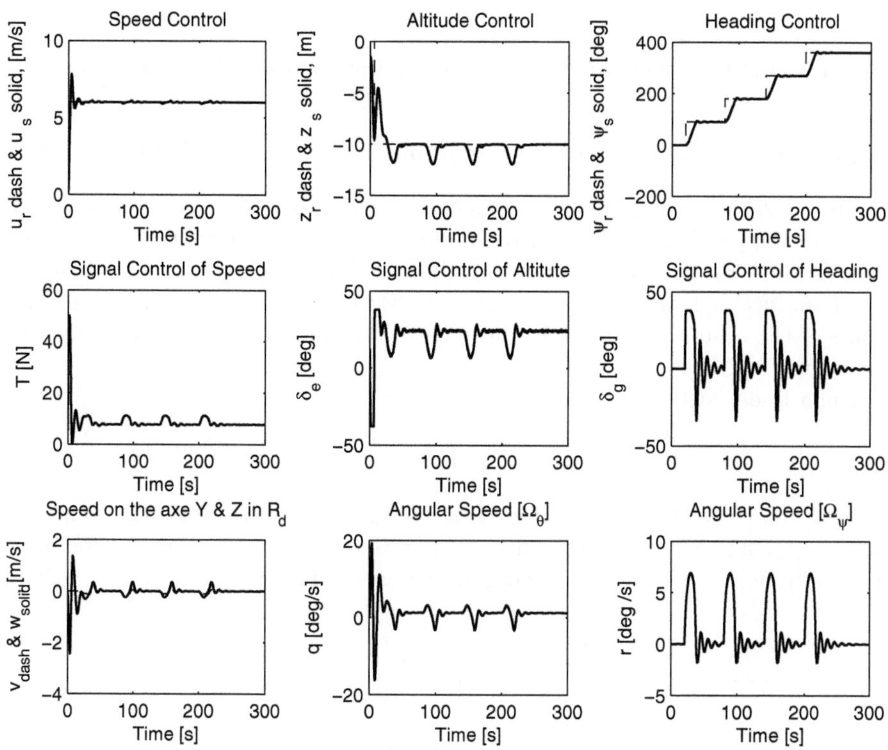

Fig. 6.17. Results of the GPC controllers

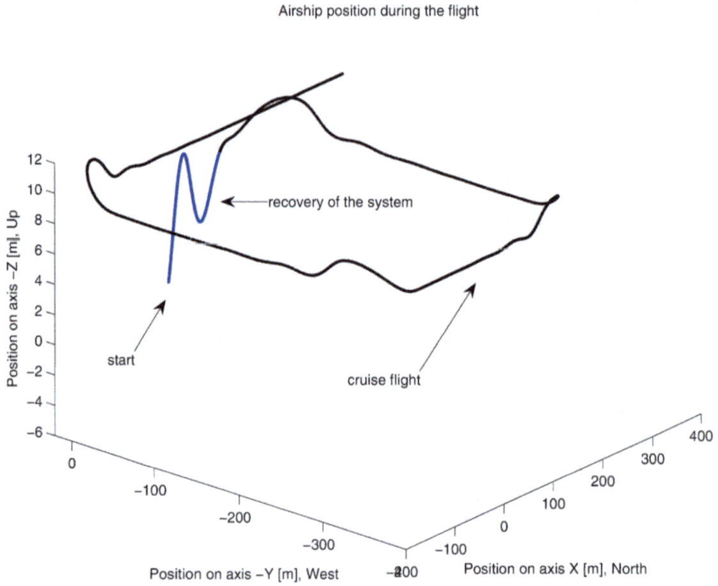

Fig. 6.18. Motion resulting from the application of the GPC controllers

6.4.3 Non-linear Control by Extended Linearization (ELC)

The linearization of non-linear system equations on equilibrium points is a very useful technique applied by control engineers. Recent control techniques have been formulated according to local or global validity. When a linearized model is not maintained, different linearized models corresponding to the operation points of the system are needed. These set of linearized models is called tangent linear model (TLM).

A non-linear system can be described by:

$$\dot{x} = f(x, u) \tag{6.43}$$

where f is a vector field parametrized by the input u and the states x. Using the Taylor approximation and considering that the system works on an operation or balance point (x_0, u_0), the approximated linearized function f can be written as:

$$f(x, u) = f(x_0, u_0) + \frac{\partial f}{\partial x}(x_0, u_0)(x - x_0) + \frac{\partial f}{\partial u}(x_0, u_0)(u - u_0) + \varsigma(x - x_0, u - u_0) \tag{6.44}$$

If the higher power terms are negligible, the equation (6.44) becomes:

$$\delta\dot{x} = \frac{\partial f}{\partial x}(x_0, u_0)\delta x + \frac{\partial f}{\partial u}(x_0, u_0)\delta u \tag{6.45}$$

This is a linear representation of equation (6.43) and the set of linearized models can be represented by:

$$\delta \dot{x} = F(x_0, u_0)\delta x + G(x_0, u_0)\delta u \tag{6.46}$$

Considering equation (6.46) and feedback theory with pole placement technique, a local control law can be formulated:

$$\delta u = -K(x_0, u_0)\delta x + \delta v \tag{6.47}$$

Integrating this control local law, it is possible to find a non-linear control law in all the system workspace [21]. Finally, this law can be expressed by:

$$u = \Omega(x, v) \tag{6.48}$$

Note that this law is valid only with small variations of the reference signal. The scheme that represents the nonlinear control by extended linearization with an anti-windup is presented in Fig. 6.19.

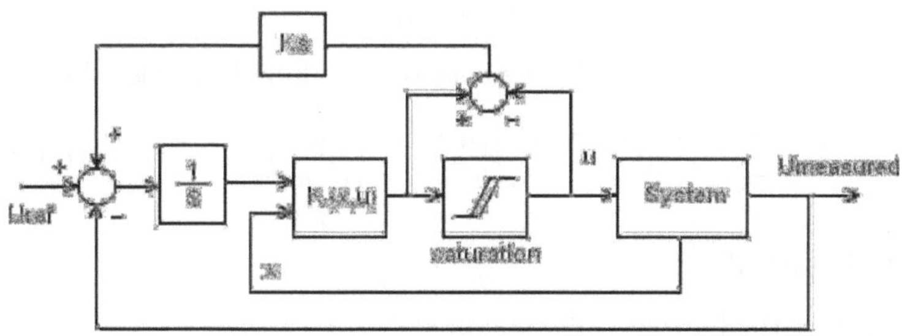

Fig. 6.19. Extended linearization regulator scheme

To begin the calculation procedure of the control law by extended linearization method, it is necessary to guarantee that the control model depends only on the desired state variable and satisfy the local governability $\frac{\partial f}{\partial u}(x_0, u_0) \neq 0$. The following developments require both conditions to be true.

Speed Controller

The speed dynamics can be represented by:

$$m\dot{u} = F_m \cos\mu + \frac{1}{2}\rho u^2 S_{ref} C_{Tt} \tag{6.49}$$

where u is the state variable and F_m is the control input. The balance point $(du/dt = 0)$ is given by:

$$u_0^2 = \frac{2F_{m0}\cos\mu}{\rho S_{ref} C_{Tt}} \tag{6.50}$$

The TLM model can be expressed as:

$$\begin{aligned} \delta\dot{u} &= \tfrac{\rho S_{ref} C_{Tt} u_o}{m}\delta u + \tfrac{\cos\mu}{m}\delta F_m \\ \delta\dot{u} &= a(u_o)\delta u + b(u_o)\delta F_m \end{aligned} \tag{6.51}$$

If the desired system performance is represented by:

$$s^2 + 2\xi w_o s + w_o^2 \tag{6.52}$$

the control law is:

$$\delta F_m = -K(u_0, F_{m0})\delta u + \int(\delta u_{ref} - \delta u)\mathrm{d}t \tag{6.53}$$

where δu_{ref} is the speed reference, and considering the feedback theory with pole placement technique, it is possible to formulate the characteristic polynomial of the closed-loop system and then, considering equation (6.52), the coefficient values can be determined.

$$\begin{aligned} k(u_o) &= \tfrac{-2\xi w_o m - \rho S_{ref} C_{Tt} u_o}{\cos\mu} m \\ k_a(u_o) &= \tfrac{m w_o^2}{\cos\mu} \end{aligned} \tag{6.54}$$

According to equation (6.48), the control law after the integration is:

$$F_m(t) = \frac{-2\xi w_o m u(t)}{\cos\mu} - \frac{1}{2}\frac{\rho S_{ref} C_{Tt} u^2(t)}{\cos\mu} + \frac{m w_o^2}{\cos\mu}\int(u_{ref}(t) - u(t))\mathrm{d}t \tag{6.55}$$

Altitude Controller

Let us consider the following altitude model

$$\begin{aligned} \dot{z} &= -u\sin\theta \\ \dot{\theta} &= q \\ \dot{q} &= k_2 q + k_1\theta + k_3\delta_e \end{aligned} \tag{6.56}$$

working on the operation point

$$\begin{aligned} \delta_{e0} &= 0 \\ \theta_0 &= 0 \\ q_0 &= 0 \end{aligned} \tag{6.57}$$

The TLM model becomes:

$$\begin{aligned} \delta\dot{z} &= -u\cos\theta_0\delta\theta \\ \delta\dot{\psi} &= \delta q \\ \delta\dot{q} &= k_2\delta q + k_1\delta\theta + k_3\delta\delta_e \end{aligned} \tag{6.58}$$

If the control law is expressed as:

$$\delta\delta_e = -K(z_0, \theta_0, q_0, \delta_{e0})\delta x + \int(\delta z_{ref} - \delta z)\mathrm{d}t \tag{6.59}$$

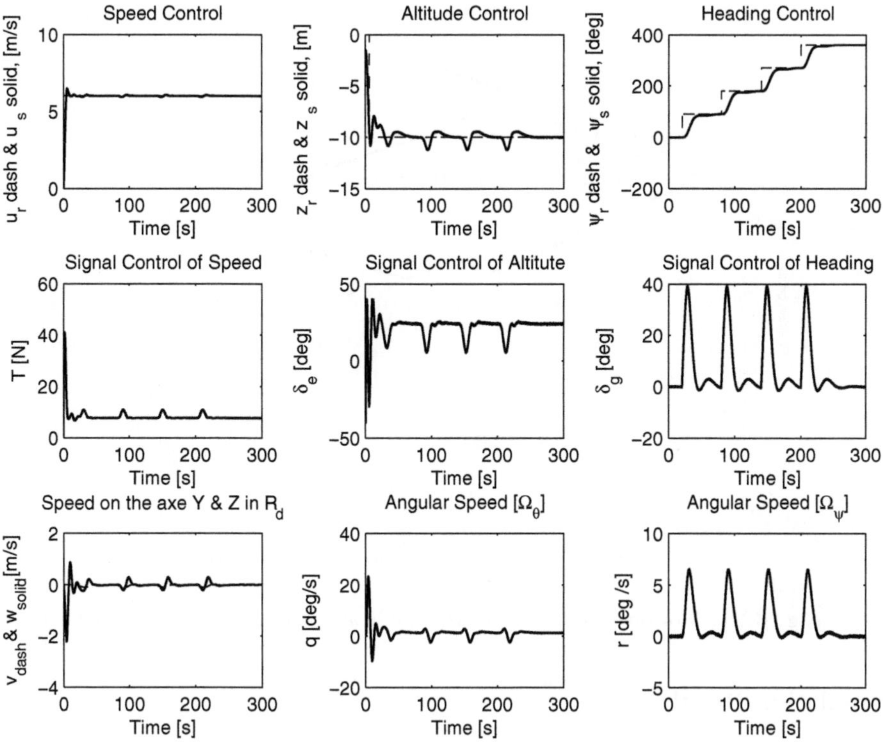

Fig. 6.20. Results of the ELC controllers

where δx is the variation of the state vector, $K = [k_z, k_\theta, k_q]^T$, and the desired performance represented by

$$(s^2 + 2\zeta w_n s + w_n^2)(s + 1/\tau_1)(s + 1/\tau_2) \tag{6.60}$$

the control law in closed-loop and after the integration becomes:

$$\begin{aligned} \delta_e = & \tfrac{w_n^2}{-u\cos\theta k_3 \tau_1 \tau_2} \int (z_{ref} - z)\mathrm{d}t - \tfrac{\tau_2 + \tau_1 + 2\zeta w_n}{\tau_1 \tau_2 k_3} q + \\ & - \tfrac{1 + 2\zeta w_n(\tau_1 \tau_2) + \tau_1 \tau_2 (w_n^2 + k_1)}{\tau_1 \tau_2 k_3} \theta + \tfrac{2\zeta w_n + \tau_1 + \tau_2}{\tau_1 \tau_2 k_3 u \cos\theta} z \end{aligned} \tag{6.61}$$

Heading Controller

Similarly to the altitude controller and with a desired performance $(s^2 + 2\zeta w_n s + w_n^2)(\tau s + 1)$, the heading controller can be expressed as:

$$\delta_g = \frac{w_n^2}{k_1 \tau} \int (\psi_{ref} - \psi)\mathrm{d}t - \frac{2\xi w_n \tau + 1}{k_1 \tau} r - \frac{w_n^2 \tau + 2\xi w_n}{k_1 \tau} \psi \tag{6.62}$$

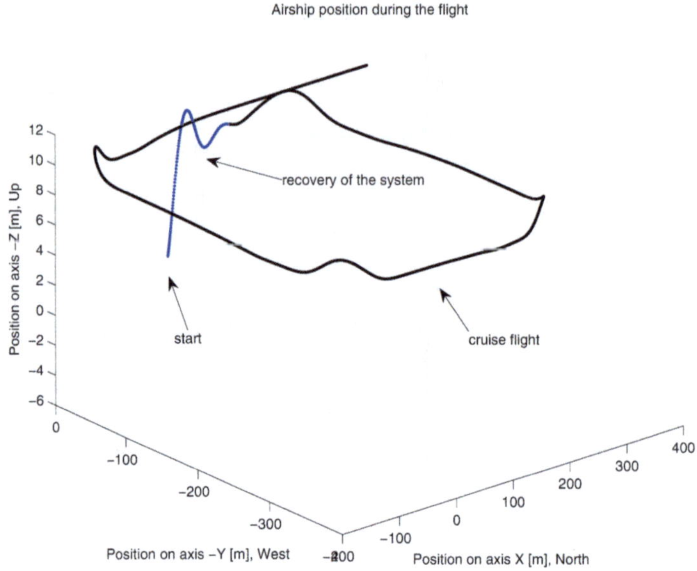

Fig. 6.21. Motion resulting from the application of the ELC controllers

Simulation Results

Figures 6.20 and 6.21 show the results of the ELC controller, with the same airship model and references as in Sect. 6.4.1.

In general it is possible to see that the controllers follow the references (speed, altitude and heading controls), even though there are couplings between the controlled states. The main difference in the controllers' performance is in the control signal that they generate. The GPC controllers try to minimize the pursuit error and the control signal energy, and consequently the response time is smaller (see Fig. 6.17). The PID and nonlinear controls have similar results with smaller control signals (see Figs. 6.14 and 6.20). Nevertheless, an advantage of the nonlinear controllers is to have the possibility of adapting automatically the parameters with the system operating point.

6.4.4 Experimental Results

Test with UrAn

The speed and altitude controllers have been tested at the UrAn airship, under a cruise flight condition. UrAn is a 28 m^3 airship built by the MiniZep company, and is the support for research held at the University of De los Andes (Bogotá-Colombia). It is equipped with a mini computer that communicates with the ground via a Wifi link, an RF receiver interconnected with the actuators, the computer and the ground remote control, a GPS, a barometric altimeter, an inertial sensor that provides the airship attitude and heading and a home-made anemometer.

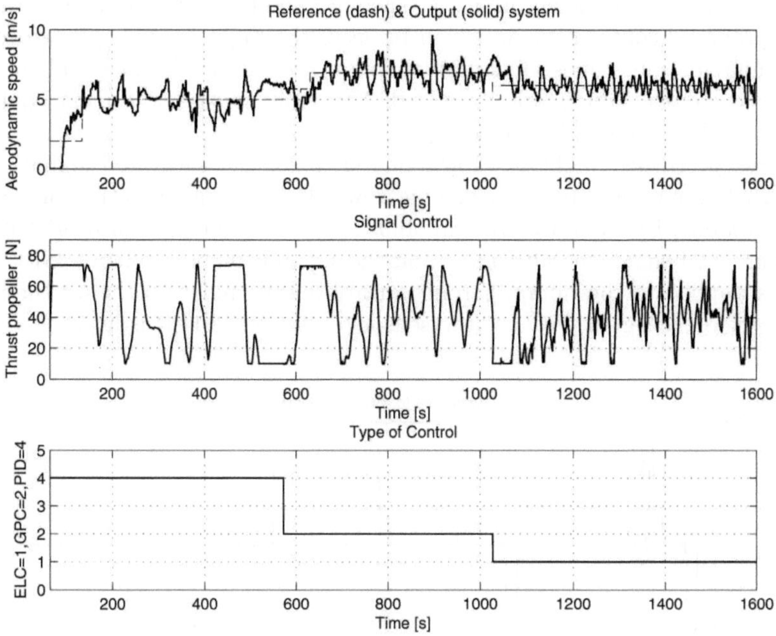

Fig. 6.22. Speed control on real system (UrAn), first result

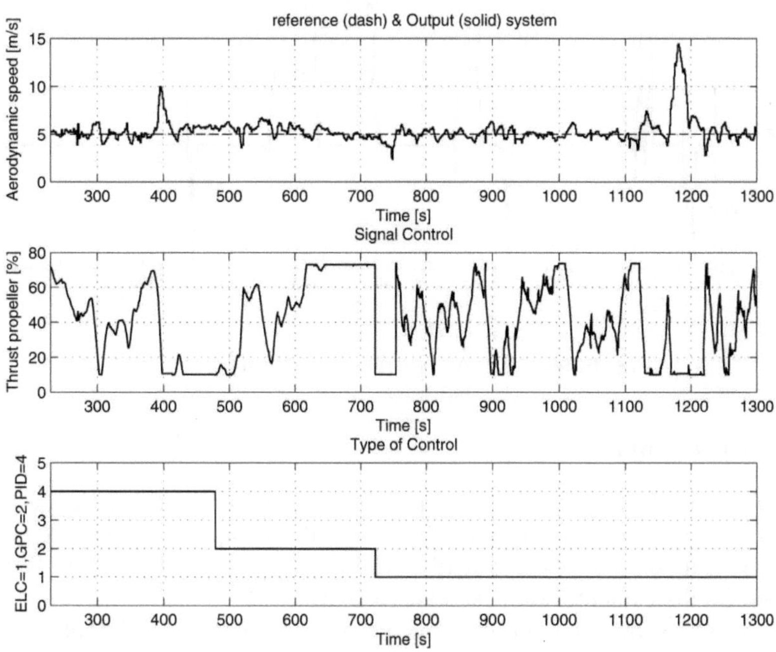

Fig. 6.23. Speed control on real system (UrAn), second result

Figure 6.22 shows the aerodynamic speed v_a and the signal control T with the three proposed controllers. As it can be seen, the flight is quite perturbed, the system having sometimes difficulties to follow the speed reference, and the some saturated control signals are generated. Nevertheless, the non-linear controller is the one that behaves the best.

A less perturbed flight is shown in Fig. 6.23, the saturated signals being generated when the airship is moving towards the wind direction or against. Unfortunately, UrAn and Karma airships are not equipped with wind sensors.

Figure 6.24 shows the altitude z and the signal control δ_e when a GPC regulator is applied to the blimp. Strong control inputs maintain the altitude close to the reference signal, because the sensor resolution (± 1 m) is low.

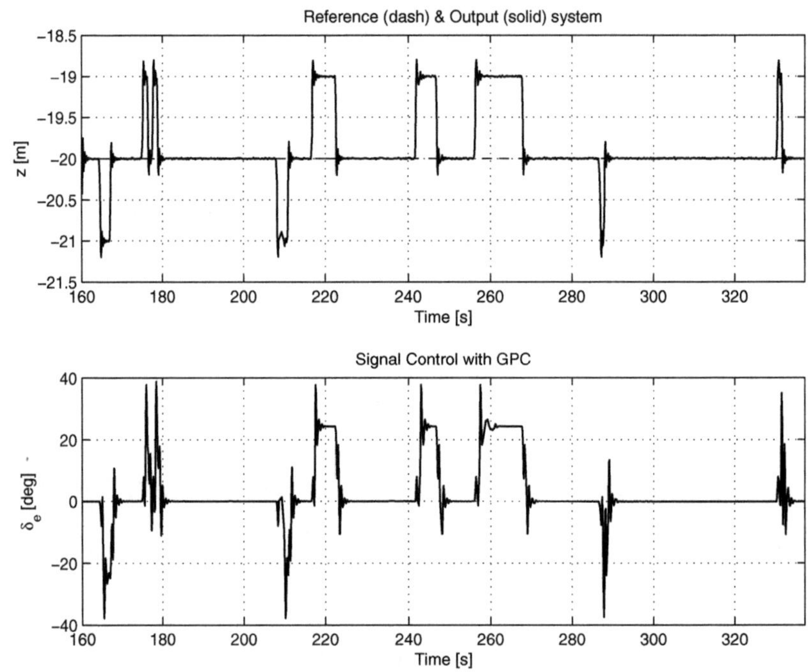

Fig. 6.24. Altitude control on real system (UrAn)

Test with Karma

The speed and heading controllers have been tested on-board with the Karma airship. The Karma hull was made by the Zodiac company, its volume is 18 m^3 and is the support for research held at the LAAS/CNRS laboratory (Toulouse-France). Its instrumentation is similar to UrAn's.

Figure 6.25 shows the aerodynamic speed v_a and the signal control T with two proposed controllers (PID and GPC). As it can be seen, the control signal

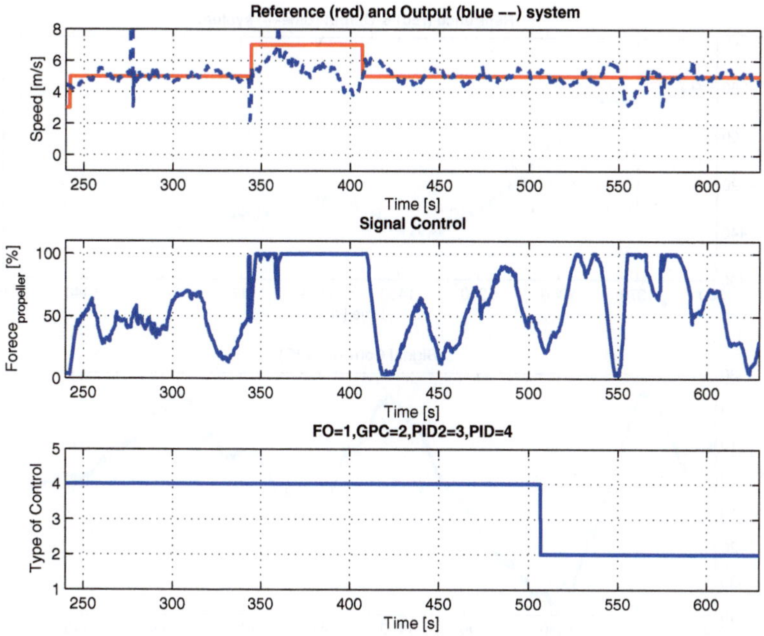

Fig. 6.25. Speed control on real system (Karma)

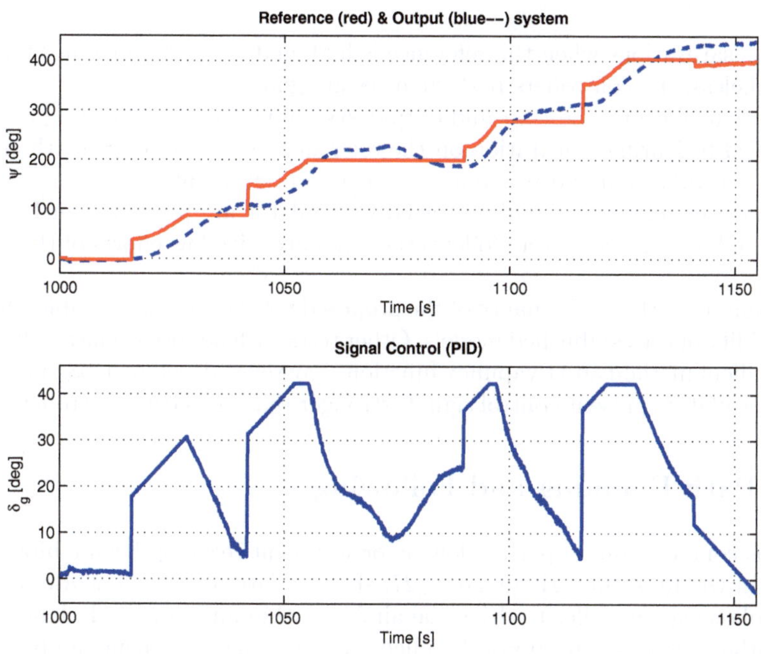

Fig. 6.26. Heading control on real system (Karma), first result

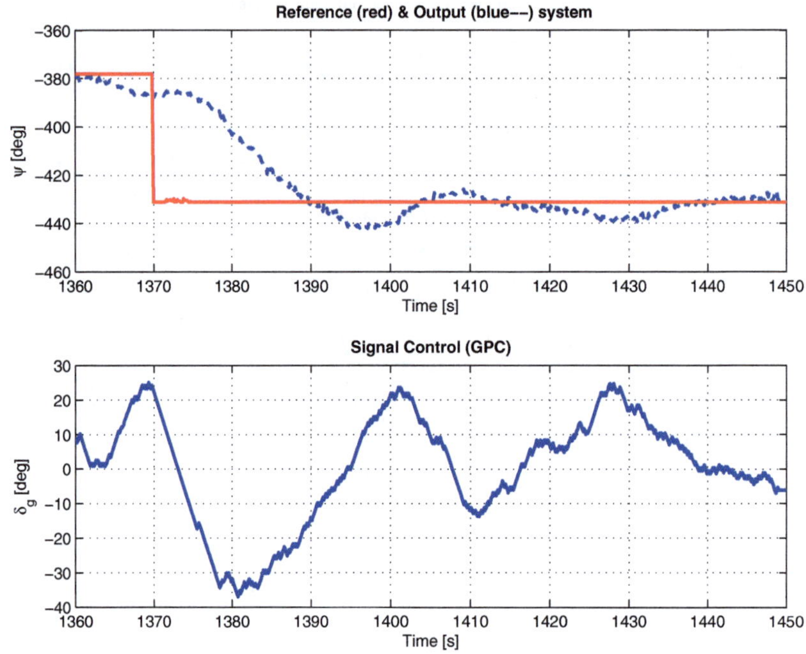

Fig. 6.27. Heading control on real system (Karma), second result

presents saturations when the reference is high and the system cannot follow it. Nevertheless, the controllers performances are good.

Figure 6.26 shows the ψ_{ref} and output system when a PID heading regulator is applied to Karma. Saturation on the reference signals assure smooth changes on the control signal and proximity to the operation point.

Change in reference signal and a GPC regulator are presented in Fig. 6.27. As it can be seen, the system follows the reference with low efforts of the control signal.

In summary, the performance of the proposed regulators is acceptable, showing the validity of the established models. Other tests with adaptive controls based on the gradient method and Lyapunov functions have been developed [38]: in general, these regulators present control signals stronger than the GPC controllers.

6.5 Path Planning and Following

Various contribution on path following or waypoint reaching for airships can be found in the literature. Hygounenc [27] shows a lateral control strategy with a path following controller to drive the airship within an horizontal plane. In [20] the authors propose an approach called "orienteering": the control objective is defined in terms of a waypoint to reach, rather than in terms of a trajectory to follow. Similar work is presented in [36], where the approach in order to obtain

ψ_{ref} is expressed in terms of the deviation from a nth waypoint. When the distance between the airship and the nth point is less than 20 m, the mission planner treats the nth waypoint as reached and the $(n+1)^{th}$ waypoint is set as the new target point. In [37], the authors present an approach in which visual signals are used to achieve a road following task – similar work can be found in [43], in the context of an indoor vision-guided blimp. In [4], the authors present two approaches to ensure the guidance of an airship: one is based on a H_∞ control technique and the other based on a PI control.

Fewer work tackle the motion planning problem. [31] describes some randomized motion planning algorithm, considering systems with both kinematic and dynamic constraints. A work based on the shortest paths for the lateral navigation of an autonomous unmanned airship is presented in [24]; the idea is the characterization of the shortest paths taking into account the dynamics and actuators limitations of the airship.

We consider here the problems of determining an appropriate trajectory within the 2D plane to reach a given goal, and of ensuring that the trajectory is properly followed. We consider the case where the airship is evolving in cruise flight, with the equilibrium conditions mentioned in Sect. 6.2: the system dynamics can be linearized.

6.5.1 Path Planning

If the longitudinal dynamics are stabilized to the steady values u_{ref} and z_{ref}, considering that the skid angle β remains negligible and the control directly acts on $\dot{\psi}$, the lateral airship's dynamics can be described by the system:

$$\begin{aligned} \dot{x} &= u \cos \psi \\ \dot{y} &= u \sin \psi \\ \dot{\psi} &= \delta_g / \rho \end{aligned} \tag{6.63}$$

where δ_g is the control to navigate in the 2D horizontal plane. This model presents two kinematic constraints: one that forces the vehicle to move tangentially to its main axis, and a second constraint due to the bound on the rudder angle prevents the blimp from turning with a radius of curvature lower than a given threshold ρ. This kinematics model was studied by Dubins [18], who considered the problem of characterizing the shortest paths for a particle moving forward with a constant velocity. In [9], Souères presents the same model and provides solutions with optimal control, the shortest path problem being equivalent to the minimum-time when the module of linear velocity is kept constant.

Rewriting the system (6.63), $\dot{x} = f(x)u + g(x)\delta_g$, it is possible to show that it is controllable. The main difficulty that arises is the under-actuation of the system, and another one is the coupling between position and direction, known as the non-holonomic constraint $\dot{x}\sin\psi + \dot{y}\cos\psi = 0$.

The problem of steering the system (6.63) from an initial configuration $\zeta_i(x)$ to a final configuration $\zeta_f(x)$, is the trajectory generation problem, that can be formulated as the minimization of $J(\delta_g)$:

$$J(\delta_g) = \int_{t_0}^{t_f} L(x(t,\delta_g), \delta_g(t))\mathrm{d}t \tag{6.64}$$

under the constraints

$$\begin{aligned} \dot{x} &= f(x)u + g(x)\delta_g \\ x(t_0) &= x_0 \\ x(t_f) &= x_f \\ \delta_{g_{min}} &\leq \delta_g \leq \delta_{g_{max}} \end{aligned} \tag{6.65}$$

Then, the problem is to find the admissible control $\delta_g \in \mathbb{R}^m$ that minimizes the time to reach the final state x_f from the initial state x_0. To solve this problem, we apply the Pontryagin's Maximum Principle (PMP) to obtain an optimal trajectory $x^*(t)$ defined on $[0,T]$ with a δ_g^* time-optimal reference control. The Hamiltonian H is defined by:

$$H = \lambda_1 u \cos\psi + \lambda_2 u \sin\psi + \lambda_3 \delta_g \tag{6.66}$$

So we have the adjoint vector, $\lambda : [0,T] \in \mathbb{R}^3$, satisfying $\dot{\lambda} = -\frac{\partial H}{\partial x}$ and $\lambda(t) \neq 0$ for every $t \in [0,T]$, and if x^*, λ^* and δ_g^* verify these conditions, then we have an extremal solution.

The extremal control can have different values, $\delta_g = \delta_{g_{max}}$, $\delta_g = \delta_{g_{min}}$ and $\delta_g = 0$, they give circular (to the right and to the left), and straight trajectories, respectively. This type of controller is called "bang-zero-bang control". The geometric approach allows to complete the total solution (see [9]). It uses the movement symmetry properties of the blimp and divides the state space, such that only six families of curves can describe the airship's movement within the 2D plane. The calculation of these trajectories is very fast and the trajectory that has the smaller distance is the optimal one.

Note that the control δ_g cannot vary continuously along a given trajectory ($\delta_g \approx$ angular speed $\dot{\psi}$). In order to make the angular speed r a variable of the system, two dynamic extensions are proposed by controlling the variation of the angular speed $\dot{r}$ instead of the angular velocity. The first extension is controlling by δ_g and the other one is controlling by $-k_{psi}r + \delta_g$.

First Dynamic Extension

If the variation of the angular speed $\dot{r}$ is directly controlled by δ_g, the new system is:

$$\begin{aligned} \dot{x} &= u\cos\psi \\ \dot{y} &= u\sin\psi \\ \dot{\psi} &= r \\ \dot{r} &= k_g \delta_g \end{aligned} \tag{6.67}$$

where k_g is a constant depending on the system dynamics. Assuming that the system (6.67) is controllable and considering that the cost function to be minimized is an energy criterion (similar to the shortest path problem when the velocity is kept constant):

$$J(\delta_g) = \int_{t_0}^{t_f} < \delta_g(t), \delta_g(t) > \mathrm{d}t \tag{6.68}$$

an optimal control δ_g^* can be found if there is an input $\delta_g(t)$ for every $t \in [0, T]$ which minimizes the cost function J. So the Hamiltonian for the system (6.67) becomes:

$$H = \frac{1}{2}(u^2 + \delta_g^2) + \lambda_1 u \cos\psi + \lambda_2 u \sin\psi + \lambda_3 ur + \lambda_4 \delta_g$$

Second Dynamic Extension

If the variation of the angular speed $\dot{r}$ is directly controlled by $-k_{psi}r + \delta_g$ (similar behavior to a first order system), the system can be written as:

$$\begin{aligned} \dot{x} &= u\cos\psi \\ \dot{y} &= u\sin\psi \\ \dot{\psi} &= r \\ \dot{r} &= -k_r r + \delta_g \end{aligned} \tag{6.69}$$

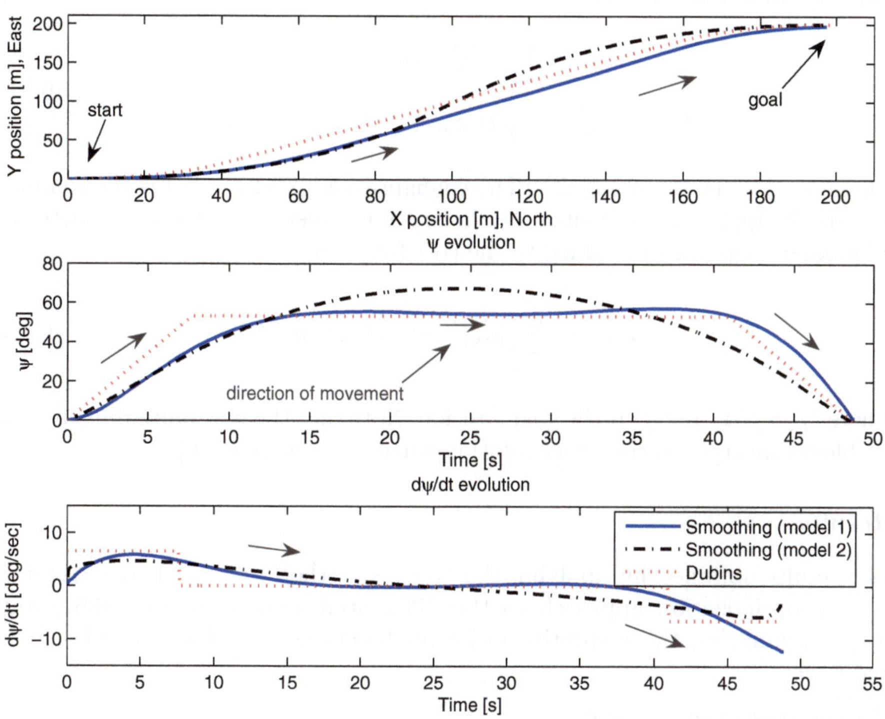

Fig. 6.28. Trajectories generated by systems (6.63) (Dubins), (6.67) (smoothing model 1) and (6.69) (smoothing model 2), for an initial configuration $\zeta_i = [0, 0, 0]$ and a final configuration $\zeta_f = [250, -250, 0]$

where k_r is a constant depending on the system dynamics. Similarly to above, there is an optimal control u^* that steers the system (6.69) from the starting point $\zeta_i(x)$ to the final point $\zeta_f(x)$ and the Hamiltonian H can be expressed by:

$$H = \frac{1}{2}(u^2 + \delta_g^2) + \lambda_1 u \cos\psi + \lambda_2 u \sin\psi + \lambda_3 ur + \lambda_4(\delta_g + uk_r r)$$

This class of systems gives non-convex problems and in general it is difficult to find a solution. One possibility to solve it is using numerical methods: the technique which we describe here was developed by Fernandes et al. [10].

The Fernandes' method: Let us consider an input control $\delta_g \in L_2([0,T])$ and denoting by $\{\mathrm{e}_k\}_{k=1}^{\infty}$ an orthonormal basis for $L_2([0,T])$[3], the continuous control law δ_g can be expressed as:

$$\delta_g = \sum_{k=1}^{\infty}(\alpha_k \mathrm{e}_k^{\mathrm{i}\frac{2k\pi t}{T}} + \beta_k \mathrm{e}_k^{-\mathrm{i}\frac{2k\pi t}{T}}) \tag{6.70}$$

and with the Ritz Approximation[4], the function δ_g can be approximated by truncating its expansion up to some rank N. The new control law and the objective function becomes:

$$\begin{aligned} \delta_g &= \textstyle\sum_{k=1}^{N} \alpha_k \mathrm{e}_k \\ J(\delta_g) &= \textstyle\int_{t_0}^{t_f} < \delta_g(t), \delta_g(t) > \mathrm{d}t \simeq \sum_{k=1}^{N} |\alpha_k|^2 \end{aligned} \tag{6.71}$$

where $\alpha = \alpha_1, \alpha_2, \ldots, \alpha_N \in \mathbb{R}^N$. The configuration $\zeta_f(x) = \zeta(T)$ is the solution at time T applying the control δ_g. In order to steer the system to $\zeta_f(x)$, an additional term must be added to the cost function,

$$J(\alpha) = \sum_{k=1}^{N} |\alpha_k|^2 + \gamma||f(\alpha) - q_f||^2 \tag{6.72}$$

where $\zeta_f = f(\alpha)$ and q_f is the goal position. Note that the new finite-dimensional problem converges to the exact solution when $N \to \infty$ (see [10]).

Results

The results obtained by applying the Souères method and the Fernandes' numerical optimization approach to the differential systems (6.63), (6.67) and (6.69) minimizing the respective objective functions, are shown in Figs. 6.28 and 6.29.

---

[3] L_2 denotes a Hilbert space.

[4] The Ritz Approximation Theory approximates the solution using solutions of some finite-dimensional problems.

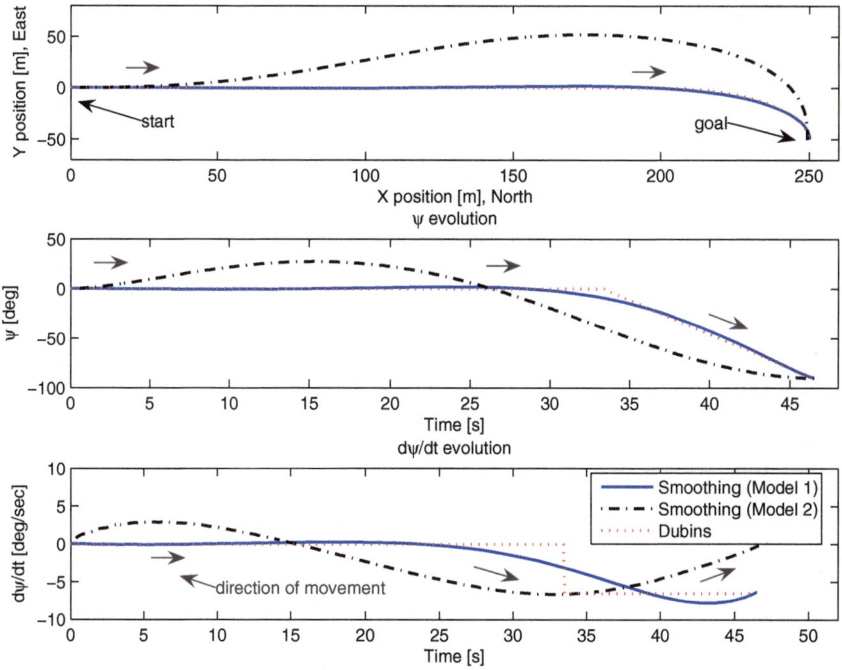

Fig. 6.29. Trajectories generated by systems (6.63) (Dubins), (6.67) (smoothing model 1) and (6.69) (smoothing model 2), for an initial configuration $\zeta_i = [0, 0, 0]$ and a final configuration $\zeta_f = [250, -250, -90]$

6.5.2 Path Following

Once a trajectory is planned, a dedicated control to ensure the path following is required. As usually done in mobile robotics, the consideration of a Frenet frame enables to formulate the path following problem as the regulation of the lateral distance and the orientation between the airship and its projection perpendicularly to the point s on the path ξ. The error dynamics can be written as:

$$\begin{aligned} \dot{d} &= v_a \sin \psi_e \\ \dot{\psi}_e &= \dot{\psi}_d - \dot{\psi}_r \end{aligned} \tag{6.73}$$

where d is the lateral distance between CV[5] and the frame R_r and ψ_e is the angular error between the blimp's orientation and the reference's orientation of the point s (Fig. 6.30).

Knowing that the dynamic of the angular speed in the cruise flight can be represented by $\dot{\psi}_d = -k1_{\psi_d}\psi_d + k2_{\psi_{ref}}\psi_{ref}$, and inspired by non-linear control (feedback linearization control), ψ_{ref} can be write as:

$$\psi_{ref} = \frac{k1_{\psi_{ref}}\psi_d + \dot{\psi}_r + v}{k2_{\psi_{ref}}} \tag{6.74}$$

[5] Volume center of the hull.

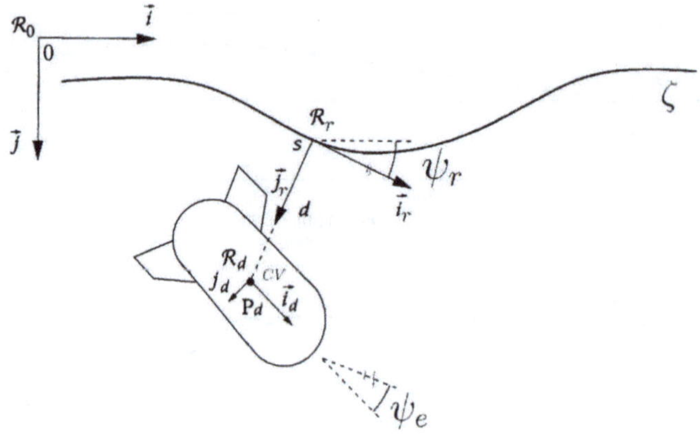

Fig. 6.30. Definition of the Frenet frame and of the parameters to regulate

Under the hypothesis that the airship is flying near the trajectory ξ (ψ_e and d remains small), the system (6.73) becomes:

$$\begin{aligned} \delta\dot{d} &= a\delta\psi_e \\ \delta\dot{\psi}_e &= \delta v \end{aligned} \tag{6.75}$$

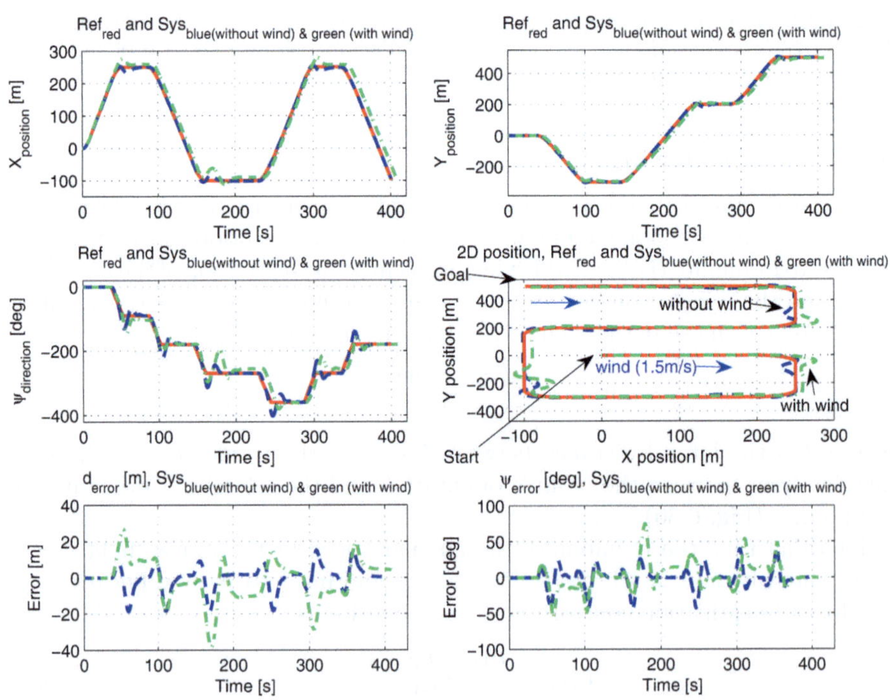

Fig. 6.31. Results of the control to follow a trajectory, Dubins model

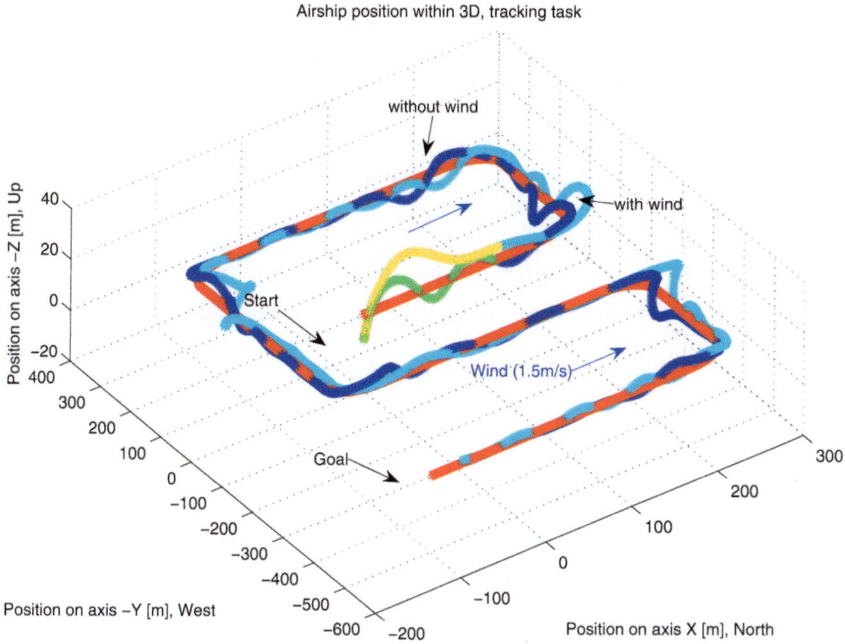

Fig. 6.32. Evolution within 3D space, Dubins model

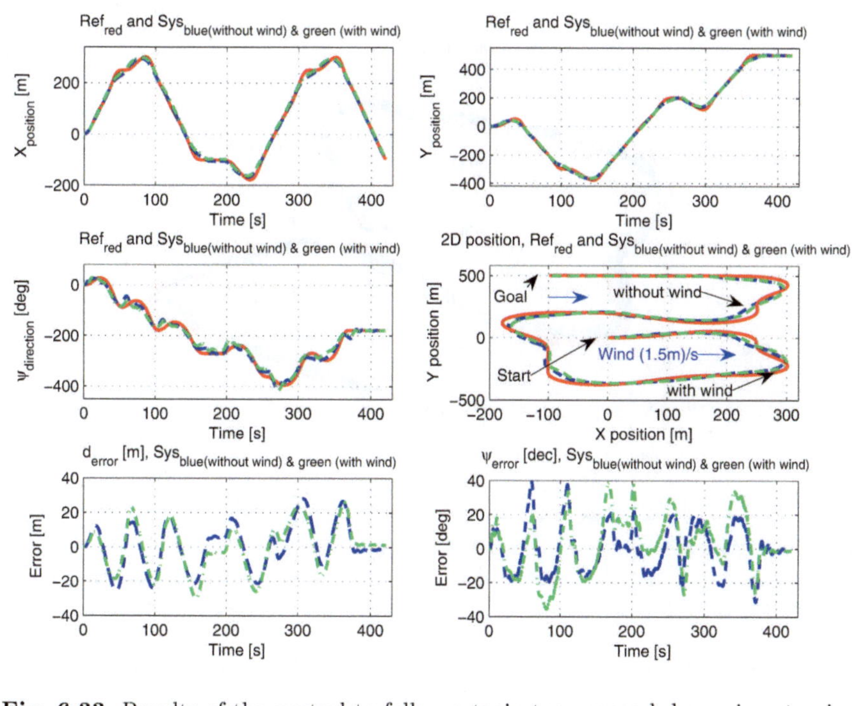

Fig. 6.33. Results of the control to follow a trajectory, second dynamic extension

where $a = v_a \cos\psi_{e0}$. The control is intended to make $d \to 0$ and $\psi_e \to 0$. For this end, a stabilizing feedback with integral control law is applied. The characteristic polynomial can be determined by:

$$\begin{vmatrix} s & -a & 0 \\ -k_d & (s - k_{\psi_e}) & k_a \\ 1 & 0 & s \end{vmatrix} = s^3 - s^2(k_{\psi_e}) - s(ak_d) + ak_a$$

Now, if the performance system in close-loop is given by $(s^3 + s^2(2\xi w_0 + \frac{1}{\tau}) + s({w_o}^2 + \frac{2\xi w_o}{\tau}) + \frac{w_o^2}{\tau})$ the values of k_{ψ_e}, k_d and k_a, are defined by the Routh-Hurwitz criterion:

$$\begin{aligned} k_{\psi_e} &= -2\xi w_o + 1/\tau \\ k_d &= -\frac{w_o^2 + 2\xi w_o/\tau}{a} \\ k_a &= \frac{w_o^2}{a\tau} \end{aligned} \tag{6.76}$$

where $d < |\frac{1}{\rho}|$ and $\psi_e \in [\frac{-\pi}{2} \frac{\pi}{2}]$. The control law for the system (6.75) is:

$$v = k_{\psi_e}\psi_e + k_d d + k_a \int (-d)\mathrm{d}t \tag{6.77}$$

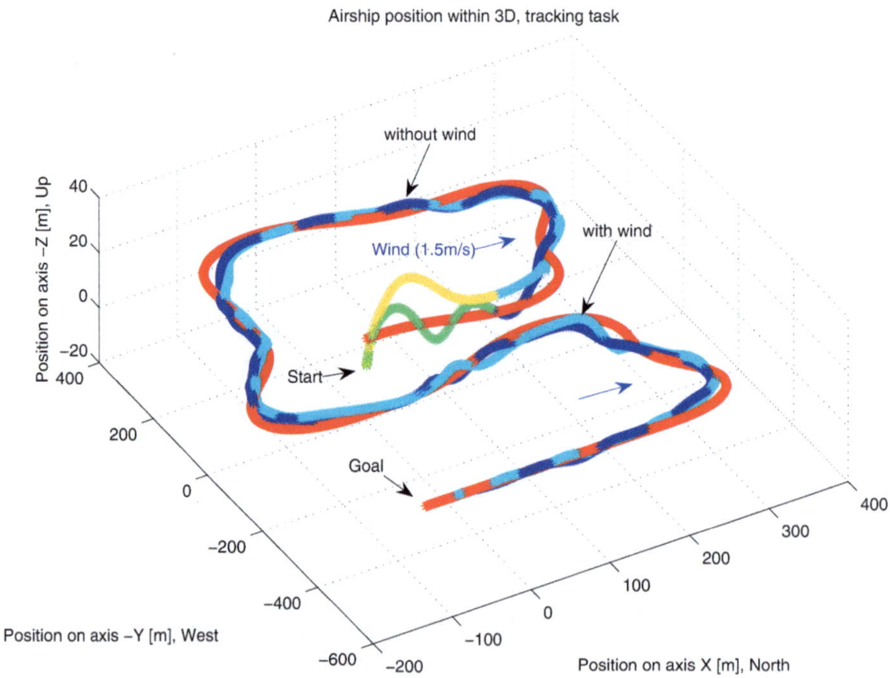

Fig. 6.34. Evolution within 3D space, second dynamic extension

Finally, the reference ψ_{ref} for the blimp heading control system that assures to follow the trajectory ξ can be written by:

$$\psi_{ref} = \frac{k1_{\psi_{ref}}\psi_d + \dot{\psi}_r + k_{\psi_e}\psi_e + k_d d + k_a \int(-d)\mathrm{d}t}{k2_{\psi_{ref}}} \tag{6.78}$$

Results

The results of path following, when the path is planned by the Dubins model, are shown in Fig. 6.31 and the airship evolution within the 3D space is presented in Fig. 6.32.

The results of our path following method and 3D airship movement, when the path is planned by the second dynamic extension model, are shown in Figs. 6.33 and 6.34.

These results compare the performance of the regulators when the airship is disturbed or not by wind. The second dynamic model has presented a trajectory more adapted to the airship dynamics and then, the lost of altitude when the blimp turns is less important.

6.6 Conclusion

This Chapter presented a complete model of an airship, which can be decoupled into simpler sub-models under the assumption that it is evolving at a stabilized speed. An identification technique based on the Unscented Kalman Filter proved to give better convergence than one based on an Extended Kalman Filter, and has been successfully applied with the airship Karma. A global control strategy that integrates a path planner, a path follower and elementary controllers has been proposed, and experimental results of cruise flight stabilization with the airships UrAn and Karma have been obtained.

This work proves that small size airships can successfully be automatically controlled, provided they evolve in cruise flight conditions. More work is to be performed when the airship is facing wind gusts and abrupt variations.

References

1. *5th Airship Association Convention, Oxford (England)*, 2004.
2. *16th Lighter-Than-Air Systems Technology Conference and Balloon Systems Conference, Arlington, Va (USA)*, 2005.
3. J. R. Azinheira. Influence of wind speed on airship dynamics. *Journal of Guidance, Control, and Dynamics*, 24(6):1116–1124, Nov–Dec 2002.
4. J. R. Azinheira, E. Carneiro de Paiva, J. Jr. G. Ramos, and S. S. Bueno. Mission path following for an autonomous unmanned airship. In *IEEE International Conference on Robotics and Automation*, San Francisco, USA, April 2000.

5. J.R. Azinheira, E. de Paiva, J. Ramos, S.S. Bueno, M. Bergerman, and S.B.V. Gomes. Extended dynamic model for AURORA robotic airship. In *14th AIAA Lighter-Than-Air Conference and Exhibition, Akron, Ohio (USA)*, July 2001.
6. D. Barnes, P. Summers, and A. Shaw. An investigation into aerobot technologies for planetary exploration. In *6th ESA Workshop on Advanced Space Technologies for Robotics and Automation, Noordwijk (The Netherlands)*, December 2000.
7. A. Bonnet and J. Luneau. Identification des coefficients aérodynamiques du dirigeable as500 du laas, rapport technique, etude hydro-aérodynamique. Technical report, LAAS/CNRS, March 2003.
8. J.H. Boschma. The devlopment progress of the U.S. Army's SASS LITE, unmanned robot airship. In *10th AIAA Lighter-Than-Air Conference and Exhibition, Scottsdale, AZ (USA)*, Sept. 1993.
9. X. N. Bui, P. Souères, J. D. Boissonnat, and J. P. Laumond. The shortest path synthesis for non-holonomic robots moving forwards. Technical report, INRIA, Report 2153, January 1994.
10. L. Gurvits C. Fernandes and Z. Li. Near-optimal nonholonomic motion planning for a system of coupled rigid bodies. *IEEE Trans. on Automatic Control*, 39(6):450–463, March 1994.
11. E.F. Camacho and C. Bordons. *Model Predictive Control in the Process Industry.* Springer, ISBN 3-540-199924-1, 1995.
12. D.W Clarke. Application of generalized predictive control to industrial processes. *IEEE Control System Magazine*, 8(2):49–55, April 1988.
13. D.W Clarke, C. Mohtadi, and P.S. Tuffs. Generalized predictive control, part i, the basic algorithm. In *Automatica*, volume 23, pages 137–148, July 1987.
14. J. L. Crassidis and F. Landis Markley. Unscented filter for spacecraft attitude estimation. In *Journal of Guidance, Control and Dynamics*, volume 26, pages 536–542, July–August 2003.
15. E. de Paiva, S. Bueno, S. Varella Gomes, M. Bergerman, and J. Ramos. A control system development environment for AURORA's semi-autonomous robotic airship. In *IEEE International Conference on Robotics and Automation, Detroit, Mi. (USA)*, May 1999.
16. E. C. de Paiva, S. S. Bueno, S. B. Gomes, Jr. G. Ramos, and M. Bergerman. A control system development environment for AURORA's semi-autonomous robotic airship. In *IEEE International Conference on Robotics and Automation*, Detroit (Michigan), USA, May 1999.
17. H. Diaz, L. Solaque, and A. Gauthier. Fuzzy identification and parallel distributed compensation for cruise flight phase of a blimp. In *V Congreso International Electrónica y Tecnologías de Avanzada - CIETA*, page 4, Pamplona, Colombia, September 2006.
18. L.E. Dubins. On curves of minimal length with a constraint on average curvature and with prescribed initial and terminal position and tangents. *American Journal of Mathematics*, 79:497–516, 1957.
19. A. Elfes, S.S. Bueno, M. Bergerman, J.G. Ramos, and S.B Varella Gomes. Project AURORA: development of an autonomous unmanned remote monitoring robotic airship. *Journal of the Brazilian Computer Society*, 4(3):70–78, April 1998.
20. A. Elfes, J. F. Montgomery, J. L. Hall, S. Joshi, J. Payne, and C. F. Bergh. Autonomous flight control for a Titan exploration aerobot. In *8th International Symposium on Artificial Intelligence, Robotics and Automation in Space*, September 2005.

21. A. J. Fossard and D. Normand-Cyrot. *Systemes non linéaries.* Instrument Society of America, North Carolina, USA, 1993.
22. S. B. Varella Gomes and J. Jr. Ramos. Airship dynamic modeling for autonomous operation. In *IEEE International Conference on Robotics and Automation*, pages 3462–3467, Leuven (Belgium), May 1998.
23. J. R. Zhenbang Gong, J. Luo, and S. Xie. A flight control and navigation system of a small size unmanned airship. In *Proc. IEEE International Conference on Mechatronics and Automation*, pages 1491–1496, Niagara Falls, Canada, July 2005.
24. S. Hima and Y. Bestaoui. Time-optimal path for lateral navigation of an autonomous underactuated airship. In *American Institute of Aeronautics and Astronomics*, August 2003.
25. S. Hima and Y. Bestaoui. Time-optimal path planning for lateral navigation of an autonomous airship. In *AIAA Guidance, Navigation, and Control Conference and Exhibit. Austin (TX), USA*, August 2003.
26. E. Hygounenc and P. Souères. A multimode control strategy for autonomous blimp. Technical report, LAAS/CNRS, March 2002.
27. E. Hygounenc and P. Souères. Lateral path following gps-based control of a small-size unmanned blimp. In *IEEE International Conference on Robotics and Automation*, Taipei, Taiwan, September 2003.
28. E. Hygpounenc and P. Souères. Automatic airship control involving backstepping techniques. Technical report, LAAS, Report 2159, October 2002.
29. M. A. Paluszek J. B. Mueller and Y. Zhao. Development of an aerodynamic model and control law design of a high altitude airship. In *AIAA 3rd Unmanned Unlimited Technical Conference*, Chicago, Il., USA, 2004.
30. S. P. Jones and J. D. DeLaurier. Aerodynamic estimation techniques for aerostats and airship. *Journal of Aircraft*, 20:120–126, 1983.
31. J. Kim and J. Ostrowski. Motion planning of aerial robot using rapidly-exploration random trees with dynamic constraints. In *IEEE International Conference on Robotics and Automation*, Taipei, Taiwan, September 2003.
32. P. Kungl, M. Schlenker, and B. Kröplin. Research and testing activities with the solar powered airship LOTTE within the scope of the airship research group at the university of stuttgart. In *14th AIAA Lighter-Than-Air Conference and Exhibition, Akron, Ohio (USA)*, July 2001.
33. I.D. Landau, R. Lozano, and M'Saad. *Adaptive Control.* Springer,June, 1998.
34. A. Moutinho and J. R. Azinheira. Stability and robustness analysis of the AURORA airship control system using dynamic inversion. In *IEEE International Conference on Robotics and Automation*, Barcelona, Spain, April 2005.
35. M. Munk, editor. *"Aerodynamics of airships", Aerodynamics Theory*, volume 6. Julius Springer, Berlin, 1936.
36. J. Rao, J. Luo, Z. Gong, Z. Jiang, and S. Xie. Design of flight control system for a robotic blimp. In *Proceedings of SPIE*, volume 6042, May 2006.
37. P. Rives and J. Azhineira. Linear structures following by an airship using vanishing point and horizon line in visual servoing scheme. In *IEEE International Conference on Robotics and Automation*, New Orleans (LA), USA, April 2005.
38. L. Solaque, A. Gauthier, and S. Lacroix. Modelamiento y control de un dirigible. In *Proceedings of the Congress of Automatic Control*, Colombia, November 2004.
39. L. B. Tuckerman. Inertia factors of ellipsoids for use in airship design. Technical report, N.A.C.A., Report 210, March 1926.

40. A. Turner. Development of a semi-autonomous control system for the uva solar airship aztec. In *3rd International Airship Convention and Exhibition, Friedrichshafen (Germany)*, July 2000.
41. T. Yamasaki and N. Goto. Identification of blimp dynamics via flight tests. *Trans. of Japan Society for Space Sciences*, 46(153):195–205, 2003.
42. T. Yamasaki and N. Goto. Implementation of robust stability augmentation systems for blimp. *Trans. of Japan Society for Space Sciences*, 46(153):155–162, 2003.
43. H. Zhang and J. Ostrowski. Visual servoing with dynamics: Control of an unmanned blimp. In *IEEE International Conference on Robotics and Automation*, Detroit (Michigan), USA, May 1999.

7

Teleoperation Tools

Nicolás Peña, Federico Cuesta, Aníbal Ollero,
Carlos Nogales, and Francisco López-Pichaco

Robotics, Vision and Control Group, University of Seville, Camino de los Descubrimientos s/n, 41092 Seville (Spain)
{nicolasp,fede,aollero,nogales,pichaco}@cartuja.us.es

Summary. This Chapter is devoted to the teleoperation of UAVs. After a short introduction, the Chapter presents the technology trends in multimodal interfaces and the use of augmented reality. Then, the hardware adaptation and the software required for the integration of a conventional remotely piloted helicopter in a multi-UAV system, namely the COMETS system, is presented. Finally, some experiments carried out in the execution of COMETS missions are described.

7.1 Introduction

Chapters 5 and 6 of this book are devoted to autonomous UAVs having on-board control systems for autonomous navigation. However, the payload and energy consumption of the on-board equipment required for the full autonomy still imposes important constraints for the small and very light UAVs. In fact most existing micro-UAVs need the teleoperation at a certain level. Particularly, operations in dynamic and cluttered environments may require the human intervention to provide reactivity to unexpected events. Moreover, it should be also noted that the ability of human guidance is today a safety requirement for the operation of UAVs.

Therefore, the application of suitable teleoperation tools is an important need for UAV systems [5]. These tools involve hardware and software technologies to provide an appropriated sensorial feedback (visual, audio, ...) to the human pilots and to generate actions at different levels from the direct guidance to waypoint and task specification. Moreover, when autonomous and remotely piloted vehicles (RPV) cooperatively perform a mission within a fleet of heterogeneous aerial vehicles, like in the COMETS project for instance, teleoperation tools are mandatory to perform the integration [7].

This Chapter is organised as follows: the next section overviews the current technology trends to ease the teleoperation of UAVs. Section 7.3 introduces the teleoperation tools developed in the COMETS project, including hardware and software issues as well as the integration of a RPV within a fleet of heterogeneous UAV and a Ground Control Centre. Section 7.4 presents experimental results of the teleoperation system in forest fires executing a mission provided by a Control Centre. The Chapter ends with conclusions and references.

A. Ollero and I. Maza (Eds.): Mult. Hetero. Unmanned Aerial Vehi., STAR 37, pp. 189–205, 2007.
springerlink.com

7.2 Technology Trends in UAV Teleoperation

A well known approach for teleoperation is to reproduce the feeling of the direct operation in the operator; that is, the aircraft is guided manually using a joystick, pedals or a similar interface to the one in the cockpit, with the operator receiving visual images from a camera mounted on the front of the vehicle. This approach has many problems that could make it not suitable for many applications [5]. The delay in both the images and sensor data sent to the operator and his commands sent back to the aircraft is one of the biggest problems. Particularly, the information provided by the images may be degraded due to bandwidth instability and transmission delays. Specific degradations include poor spatial resolution, low or not constant update rates, and delayed updating [2].

The above limitations generate many problems. For example, inconstant image update rates degrade perception of motion information that is useful for drawing attention to air traffic and to sense the aircraft behavior leading to instability of manual UAV control [3].

Another issue is the fact that the feeling of operation and responsibility is not as intense as when the pilot is on board, chairing fate with the aircraft.

The conclusion is that trying to recreate the real operation feeling on ground, relying mostly in visual information taken provided by on-board cameras, is not always the best solution and that other ways of providing data to the pilot should be used to complement or even substitute the classical ones.

The technologies that are being used to accomplish this task are mainly built around the ideas of multimodal human machine interfaces, augmented reality and pure synthetic images. In the next paragraphs a state of the art of these fields is roughly provided.

7.2.1 The Need of Multimodal Interfaces

As mentioned above, one of the problems of the physical separation between aircraft and operator is that a range of sensory cues available to the pilot of a manned vehicle is not available to the teleoperator. Rather than receiving direct sensory input from the environment and the intense feeling of presence, the UAV ground operator receives only that information provided by on-board sensors via data link from the usually very distant UAV. This approach also creates problems and the interface designers must do their best when designing how they provide the sensed information to the operator thinking in every situation, including malfunctions, and comparing with the real flight experience.

Visual data overlaid on screen is an immediate option to inform the operator of sensed data beyond that conveyed by images from the camera. For example, in the COMETS project, a set of four virtual monitors were drawn in the corners of the pilot interface and were used to provide information about the UAV position and orientation, and the mission to be accomplished. The most useful of these virtual monitors was a virtual horizon that was refreshed at a constant and fluent rate that helped the pilot to stabilize a teleoperated helicopter.

However, UAV operators may not modify their visual scanning methods to compensate for the non-recreated multisensory cues [10]. Then, in addition of visual information, other ways of alerting the operators should be used in conjunction, as for example multimodal information displays. Thus, [8] revealed that haptic alerts, conveyed via the UAV operator's joystick, could improve self-rated situation awareness during turbulent conditions in a simulated UAV approach and landing task.

Many other publications as for example [1, 11, 12] have shown advantages of multimodal interfaces. These publications have shown that audio and tactile messages can improve many aspects of flight control and overall awareness of the situation. In the COMETS project we have implemented an audio interface used by the computer to tell the pilot corrections needed in course. A set of prerecorded words are used to tell the pilot the corrections needed to achieve the desired course. This interface is also used for direct communication between the members of the ground station and had proven to be very useful.

7.2.2 Augmented Reality and Synthetic Images

The use of "augmented reality" [6] refers to the mixture of real images taken by cameras on-board with synthetically generated landmarks or objects in relation with the scene in order to increase the usefulness of the visual information provided to the viewer. The synthetic nature of the elements overlaid to the real imagery is a key advantage in the context of UAV teleoperation because of its independence from the sensors on-board. This makes possible to update immediately these elements in response to control inputs from the operator, providing rapid feedback in such a way that the operator feels a more natural response and then makes easier a fine operation. For example, in [3] a computer-generated grid of perpendicular lines was overlaid on camera image, and moved in response to camera inputs. The synthetic grid shifted in real time following the input from the operator, gives visual feedback as to the direction and magnitude of camera movement. It was experimentally tested tracking a moving ship with a simulated UAV sensor camera. As compared to a control condition with no virtual grid, augmented displays significantly improved target tracking at low camera update rates (i.e., long delays).

A more sophisticated form of display is a fully virtual system, in which terrain information is stored in databases and rendered based on GPS position. One of the main advantages of this method is that only attitude data are mandatory in the control station to operate. Then, if the imagery data channel fails, the operator can rely on the generated virtual views to keep the control. One danger that should be taken into account when designing the virtual reality engine is that, if the images are too realistic, the operators may trust too much on the data they provide, leading to cognitive tunneling and neglecting information not available within such a high imagery display, e.g., a "transponder off" aircraft [9]. Augmented displays thus present a promising method of enabling good UAV operator performance, but are very difficult to develop because of the large number of options and the fact that a failure in design may lead the operator to relay too much on virtual images or even to get confused.

Reference [4] presents an annotation-based rescue assistance system for a teleoperated unmanned helicopter with a wearable augmented reality (AR) environment. The helicopter is remotely controlled by an operator while watching an annotated view from the helicopter through a head mounted display (HMD). Virtual buildings and textual annotations assist the rescue operation indicating the position to search rapidly and intensively. Both, position and helicopter's attitude are measured by a GPS and a gyroscope.

In the COMETS teleoperation station, the payload operator's interface (whose mission was to point a camera controlling the pan & tilt) included marks overlaid. Such marks, reflecting the relative angle of the camera, were drawn at a much higher rate than the images and a spot was drawn to simulate the target to ease the operator activity.

We are now developing a whole new version of the teleoperating software and using synthetic images for not relaying too much on the visual information sent from on-board cameras and thus relaxing the need for a very robust wide band connection. In our opinion, this is possible if the UAVs have a minimum grade of automated control. It could be also the best path to follow when trying to control more than one UAV at the same time as it permits rendering the most appropriate point of view in each situation.

7.3 Teleoperated Helicopter in COMETS

This section presents the teleoperation tools developed to integrate a remotely piloted vehicle (RPV) within the COMETS system. These tools make possible that a RPV collaborates with UAVs, within an heterogeneous fleet of aerial vehicles, to execute a mission which is coordinated by a Ground Control Centre.

The COMETS Ground Control Centre is composed of three subsystems:

- The Mission Planning System (MPS) which plans and monitors the global COMETS system mission.
- The Monitoring and Control System (MCS) which monitors the state of each UAV.
- The Perception System (PS) which processes images and data sent by the UAVs for event monitoring, terrain mapping and alarm detection and localisation (see Chap. 4).

Therefore, the teleoperation station main objective is to provide the pilot with skills to improve navigation in low visibility conditions, control of payload, as well as information from the MPS and the MCS, including prescribed waypoints, dangerous areas, location of the UAVs currently flying, and any other useful information to cooperate in the mission fulfillment (see Chap. 2).

These tools requires both hardware and software elements that are discussed separately in the following.

7.3.1 Hardware Elements

The teleoperated UAV is a conventional radio controlled helicopter mechanically adapted by the company Helivision, and which have been transformed by the Robotics, Vision and Control Group (GRVC) at the University of Seville by adding sensing, perception, communication and control functions (see Fig. 7.1). The hardware of the teleoperation system is composed of both on-board and on-ground components.

Fig. 7.1. Helivision-GRVC helicopter

On-Ground Hardware

With respect to the on-ground hardware it should be remarked that the helicopter is controlled by using conventional helicopter remote control devices. Thus, RC is used to both control helicopter servos and on-board cameras payload. The ground station for teleoperation has been designed for easy transportation and deployment (see Fig. 7.2).

Fig. 7.2. Easy deployment of the teleoperation station developed in COMETS

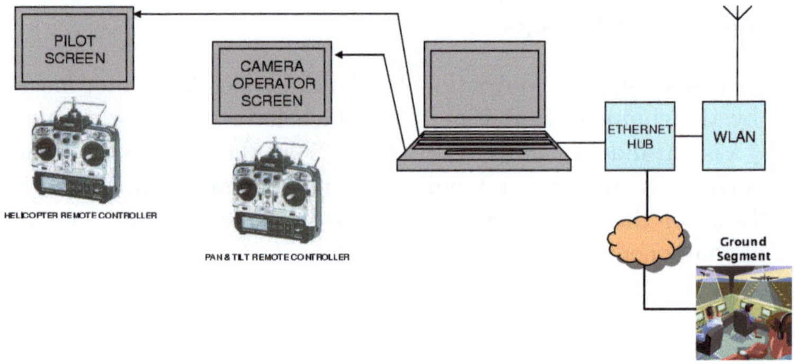

Fig. 7.3. On-ground hardware scheme

The central part (see Fig. 7.3) is a notebook computer with the teleoperation station software. In addition, two external sun readable screens can be connected for the pilot and the operator of the cameras. Section 7.3.2 will illustrate in detail the information shown on each screen. The pilot's screen is used for helping the pilot to control the helicopter providing attitude, velocity, and mission status information as well as images from the on-board camera. It is installed on the remote control of the helicopter to be more accessible for the pilot, as shown in Fig. 7.4.

Fig. 7.4. Helicopter's pilot with pilot screen on the remote control

The helicopter controller has a wireless link to communicate with the on-board hardware and an Ethernet link to connect the teleoperation station with the COMETS network (see Chaps. 2 and 3).

On-Board Hardware

The hardware on-board the helicopter includes a set of sensors and devices for data processing and data link with the ground station. Thus, the helicopter is equipped with:

- GPS.
- Attitude and velocity sensors.
- Barometric pressure sensor.
- Wireless Ethernet IEEE 802.11g access point.
- Digital video server (AXIS 2400).
- On-board microcontroller.
- Teleoperation camera.
- Visual and infrared cameras (Raytehon ControlIR 2000AS).
- Pan & tilt device.

The on-board hardware architecture is shown in Fig. 7.5. Several blocks can be distinguished in this scheme. The digital video server, for instance, is used to acquire, digitalize, and send to the ground the images from the on-board visual and infrared cameras. The on-board microcontroller is also used to acquire and manage data from other sensors. On the other hand, a visual camera and the attitude sensors provide the images which help the pilot to sense the situation visually, and angular information used for stabilization and orientation, respectively. The GPS provides the helicopter position to be visualized in the ground

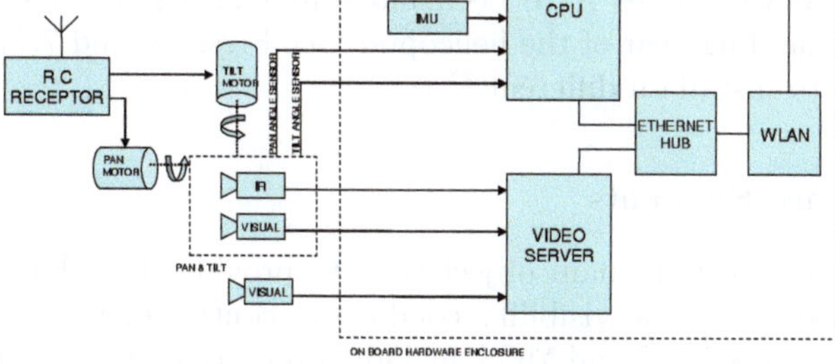

Fig. 7.5. On-board hardware architecture

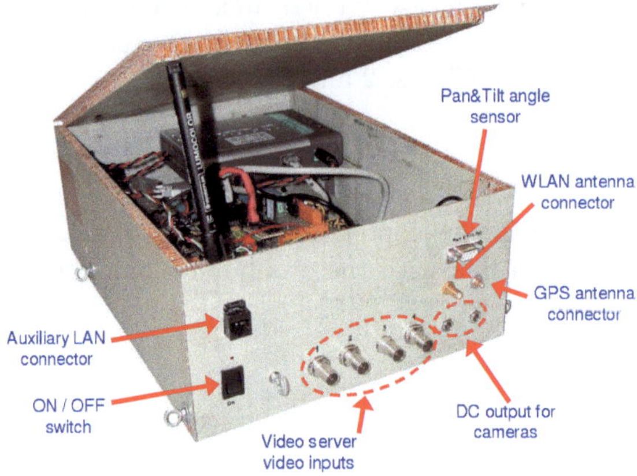

Fig. 7.6. Current on-board hardware enclosure

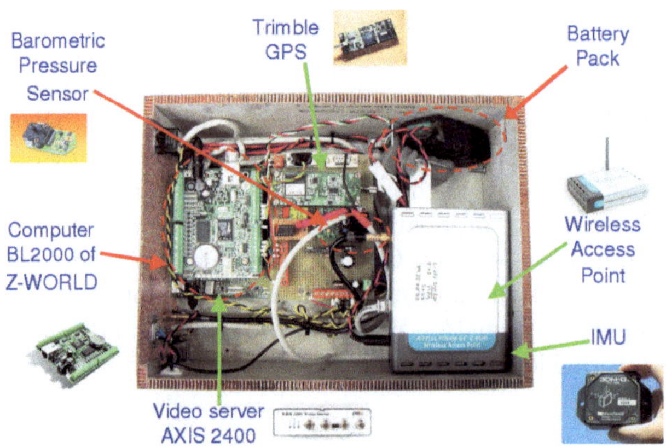

Fig. 7.7. Devices inside the current on-board hardware enclosure

control station and used for the helicopter guidance and coordination with other members of the team.

All of the above devices have been integrated in a glass fiber box that is hooked on the landing gear of the helicopter (see Figs. 7.6 and 7.7). In this way it is very easy to install on different UAVs.

7.3.2 Software Elements

The teleoperation station main objective is to provide the pilot with skills to improve navigation in low visibility conditions, control of payload, as well as information from the MPS and MCS, including prescribed waypoints, dangerous areas, location of the UAVs currently flying, and any other useful information to cooperate in the mission fulfillment (see Chap. 2).

The teleoperation station itself can be implemented by using one or more computers. Moreover, the possibility of using several displays within the same computer is also considered. This is related with the fact that two or even three operators are possible (being two the minimum). The corresponding software architecture is composed by a set of modules, implemented in a heavily multi-threaded way, performing different tasks such as UAV data sensor acquisition, visualization of UAV sensor status, configuration, realtime logging, control of the video server and communication with other components of the COMETS platform using the Communication System presented in Chap. 3.

In this way, the following information is provided to the operator in a graphical Human Machine Interface (see Fig. 7.8):

- Helicopter attitude and global position.
- GPS time.
- Telemetry from helicopter's sensors.
- Images from the on-board teleoperation cameras.
- Images from the infrared and visual on-board cameras.
- Commanded waypoints and tasks from the Control Centre.
- Deviation from the prescribed heading.
- Trajectory followed by the helicopter.
- Dangerous or forbidden areas.
- Location of the surrounding UAV fleet.
- Interface for messages interchange with the Control Centre.

Fig. 7.8. Teleoperation station screen during a fire monitoring experiment

Therefore, the teleoperation software has three main functions:

- Helicopter communication.
- Integration and communication with the COMETS system.
- Implementation of a Human Machine Interface (HMI).

The first one concerns the transmission of the on-board acquired information (digital images, GPS data, velocity, altitude, etc.) over a wireless Ethernet link by using TCP/UDP/IP protocols. A simple time saving private protocol has been implemented for data interchanged between the on-board and on-ground computers of the teleoperation system.

The second one is related to the integration in the COMETS system (see Fig. 7.9), which has two main goals. On the one hand, to distribute the information collected by the helicopter to the rest of the UAVs in the COMETS system and to the control centre. On the other hand, it provides to the pilot with valuable information about the position of the different UAVs as well as with waypoints to perform a specific mission, GPS corrections, etc. In this way, for instance, it is possible to perform cooperative perception and monitoring.

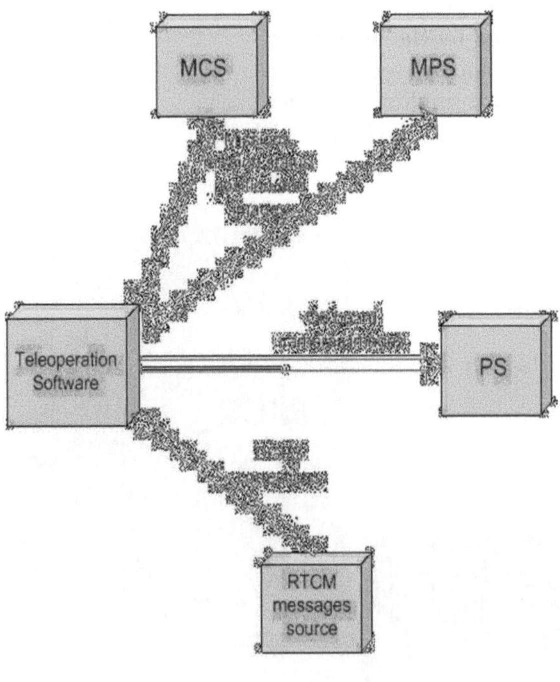

Fig. 7.9. Integration of the teleoperation software within COMETS system

Human Machine Interface (HMI)

The HMI has been designed in such a way that it is possible to use a multi-screen setup. Then, a small screen can be placed on the pilot RC whereas others can be used to display more general information or just for camera control. Figure 7.10, for instance, shows a detailed view of the instrumentation and mission tracking panel. This window includes:

- GPS data: position, altitude, velocity, heading and GPS status messages.
- Barometric altitude.
- Yaw angle, measured by the IMU sensor.
- Battery supply status.
- Helicopter trajectory.
- Waypoints to be followed.

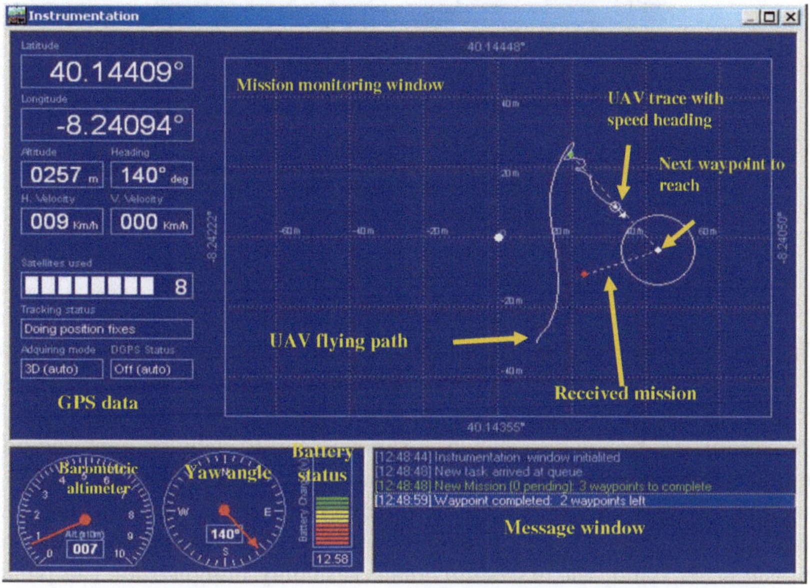

Fig. 7.10. Instrumentation and mission tracking panel

The trajectory and mission data display window, can be zoomed and/or centered to improve performance. There is also a message window for information interchange with the MCS, as well as to report on mission events. The teleoperation station shows the video sequence from the on-board cameras and overlap relevant information to help the pilot and/or the camera operator in the mission execution. There are two main camera image window modes: *Pilot Mode* and *Assistant Mode*. Visible or I/R camera images can be used in both modes.

In the *Pilot Mode* several tools are overlapped on the images as it is shown in Fig. 7.11. This window includes just the information relevant to the pilot for controlling the helicopter and to perform the mission received from a Control Center. This is usually displayed on the pilot screen installed on the remote control (see Fig. 7.4).

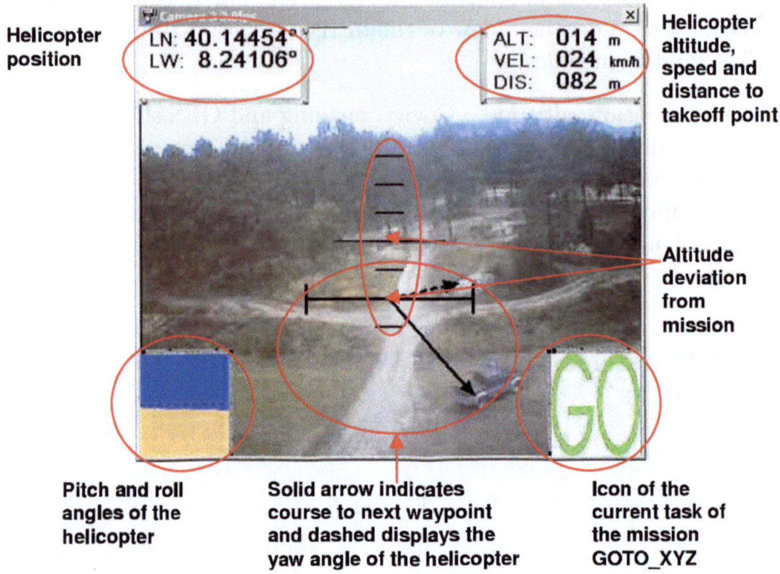

Fig. 7.11. Camera image window in *Pilot Mode*

This information includes a numerical indication of current helicopter position, altitude, velocity and distance to the take-off point, as well as a graphical description of the current helicopter pitch and roll.

Mission information is also shown in the window. It includes, centered on the screen, a graphical indication of the altitude and heading deviation from the goal, which is very useful for the pilot to reach the target location. An icon with the current task of the mission is also shown.

On the other hand, the *Assistant Mode* is mainly designed for the pan&tilt operator, as shown in Fig. 7.12. The assistant operates the pan&tilt with visible and/or IR cameras by using a conventional remote control. The current pan&tilt angles can be measured by the on-board CPU and sent to ground, and the orientation is overlapped by means of a cross and numeric values. In the current prototype of the helicopter, the pan&tilt and hardware enclosure have been installed on the landing gear and can be turn 360 degrees. In this way, during monitoring tasks for instance, it is possible to cover a wider area without changing the helicopter position or heading.

The different windows of the teleoperation station (camera, indicators, network and logging) can be accessed from setup windows. These windows are used to display the information required to monitor the state of the UAV, communication and other critical information.

A camera configuration interface allows to activate/deactivate the visualization of images from the on-board cameras as well as to setup parameters (resolution, frame rate, ...) for each channel. Moreover, pilot or assistant mode configuration can be also selected.

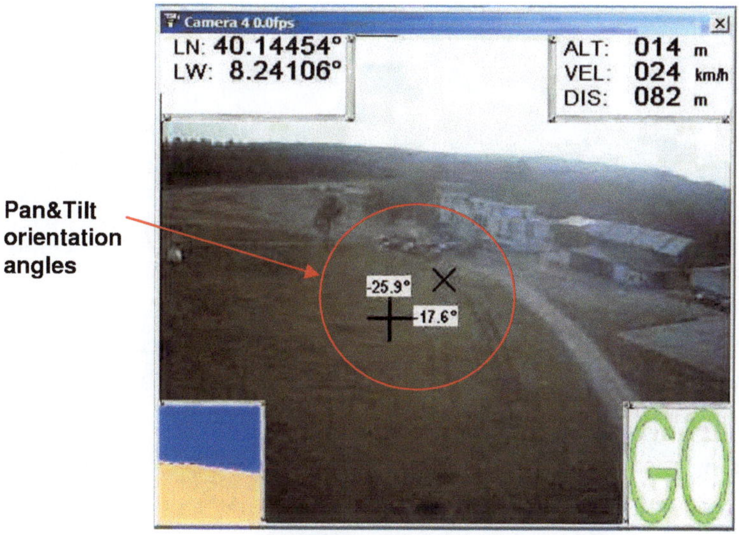

Fig. 7.12. Camera image window in *Assistant Mode*

The teleoperation system is able to automatically switch to B&W, decrease resolution and increase compression when the video stream is near to the maximum bandwidth of the WLAN. In this way, the teleoperator receives a smooth and continuous flow of images.

Finally, a network configuration window allows to change communication parameters, such as IP addresses. This window can be also used for bandwidth consumption monitoring. Thus, both image and data streams can be monitored.

7.4 Experimental Results and Mission Execution

The teleoperation tools have been intensively tested during the COMETS project experiments (see Chap. 8). In these experiments the teleoperation system performed mission tracking and fire monitoring experiments exhibiting a good overall performance [7]. Figure 7.13 shows several photographs of the teleoperated helicopter during a fire monitoring experiment in the Serra de Gestosa, near Coimbra (Portugal). A full video of the experiment can be downloaded from the COMETS web site [2].

7.4.1 Mission Execution

In the following, a simple mission execution example will be described in order to better illustrate the teleoperation tools presented in this Chapter. This mission is composed by take-off (TO), go-to (GT) and land (LD) tasks (see Chap. 2). Along the mission, text messages in the logging window report on mission events.

Fig. 7.13. Fire monitoring experiments with the teleoperated helicopter in Gestosa (Portugal) during 2003

Take-Off

The first step is the take-off task. The pilot receives the command for taking off as shows the task icon in Fig. 7.14.

Go-to Next Waypoint

Once the helicopter has taken off, the collection of waypoints is displayed on the mission panel, and the helicopter starts navigating to the waypoints, as shown in Fig. 7.15. Notice the "GO" icon on the screen. The target waypoint is painted white, highlighted by a circle. When the pilot aligns the dashed and solid arrows, the helicopter is heading for the target waypoint (see Fig. 7.15, left side).

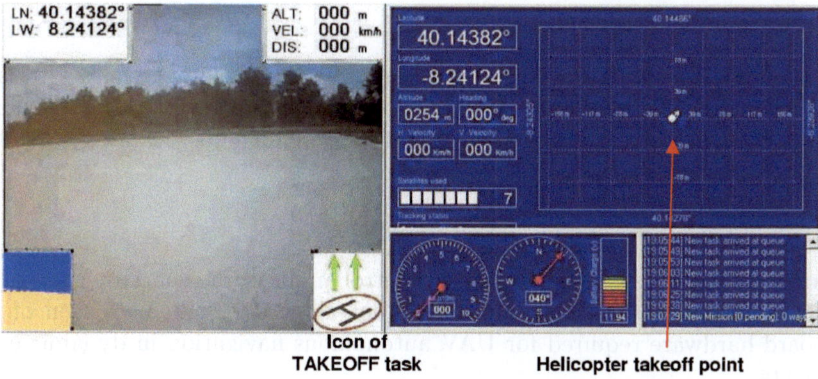

Fig. 7.14. Mission execution step 1: Take-off

Fig. 7.15. Mission execution step 2: helicopter is heading for a target waypoint

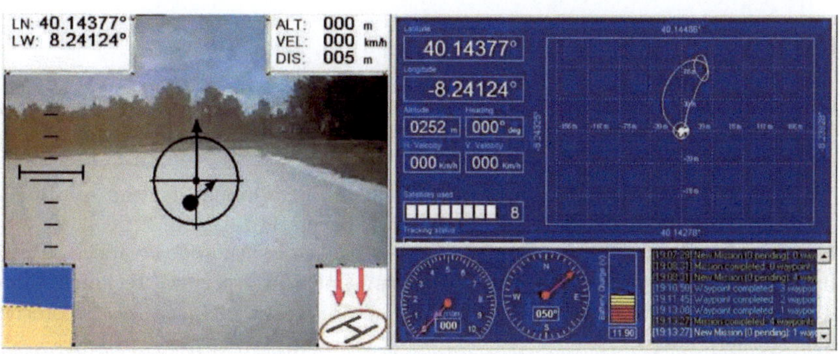

Fig. 7.16. Mission execution step 3: helicopter approaching landing point

Once the waypoint is reached its dot changes from white to green color and a new target waypoint is highlighted with a circle. This process is repeated from one waypoint to another.

Landing and Ending Mission

When the helicopter has completed the last waypoint, the next task is to land (see Fig. 7.16). The task icon changes to inform the pilot. The default landing point is the same as the takeoff point, and it is highlighted with a circle.

7.5 Conclusions

Teleoperation could be the only way to control an unmanned aerial vehicle when the payload and energy consumption do not allow the implementation of the on-board hardware required for UAV autonomous navigation in dynamic environments.

Therefore, the application of a suitable and easy to deploy teleoperation station could be the preferred approach in many applications. This approach requires hardware and software technologies to provide an appropriated sensorial feedback to the human pilots and to generate actions at different levels from the direct guidance to waypoint and task specification, as well as, integration of a remotely piloted vehicle within a fleet of heterogeneous UAVs coordinated by a Ground Control Centre.

Some of these technologies have been reviewed in this Chapter and have been illustrated by means of the teleoperated tools developed in the frame of the COMETS project.

A new teleoperation station in being developed using synthetic images for not relaying to much on the visual information sent from the on-board cameras and thus relaxing the need for a very robust wide band connection only requiring a minimum grade of automated control in the UAV. It could be also a good option to control more than one UAV at the same time as it permits rendering the most appropriate point of view in each situation.

References

1. G.L. Calhoun, M.H. Draper, H.A. Ruff, and J.V. Fontejon. Utility of a tactile display for cueing faults. In *Proceedings of the Human Factors and Ergonomics Society 46th Annual Meeting*, pages 2144–2148, 2002.
2. Van Erp J.B.F. Controlling unmanned vehicles: The human factors solution. In *RTO Meeting Proceedings 44 (RTO-MP-44)*, pages B8.1–B8.12, 2000.
3. Van Erp J.B.F. and L. Van Breda. Human factors issues and advanced interface design in maritime unmanned aerial vehicles: A project overview. Technical Report TNOreport TM-99-A004, TNO Human Factors Research Institute, Soesterberg, The Netherlands, 1999.
4. M. Koeda, Y. Matsumoto, and Y. Ogasawara. Annotation-based rescue assistance system for teleoperated unmanned helicopter with wearable augmented reality environment. In *Proceedings of the 2005 IEEE International Workshop on Safety, Security and Rescue Robotics*, Kobe, Japan, June 2005.
5. J.S. MacCarley and C.D. Wickens. Human factors implications of uavs in the national airspace. Technical Report AHFD-05-05/FAA-05-01, University of Illinois, Institute of Aviation, Aviation Human Factors Division, April 2005.

6. P. Milgram and H. Colquhoun. A taxonomy of real and virtual world display integration. In Y.O.H. Tamura, editor, *Mixed Reality: Merging Real and Virtual Worlds*, pages 5–30, Tokyo, 1999. Springer-Verlag.
7. A. Ollero, J. Alczar, F. Cuesta, F. Lpez-Pichaco, and C. Nogales. Helicopter teleoperation for aerial monitoring in the COMETS multi-UAV system. In *Proceedings of the 3rd IARP Workshop on Service, Assistive and Personal Robots (IARP)*, pages 137–142, Madrid, Spain, october 2003.
8. H.A. Ruff, M.H. Draper, L.G. Lu, M.R. Poole, and D.W. Repperger. Haptic feedback as a supplemental method of alerting UAV operators to the onset of turbulence. In *Proceedings of the IEA 2000/ HFES 2000 Congress*, pages 3.41 – 3.44, 2000.
9. L.C. Thomas and C.D. Wickens. Eye-tracking and individual differences in unexpected event detection when flying with a synthetic vision system display. In *Proceedings of the 48th Annual Meeting of the Human Factorand Ergonomics Society*, Santa Monica, CA:HFES, 2004.
10. A.P. Tvaryanas. Visual scan patterns during simulated control of an uninhabited aerial vehicle (UAV). *Aviation, Space, and Environmental Medicine*, 75(6):531–538, 2004.
11. C. D. Wickens and S. Dixon. Workload demands of remotely piloted vehicle supervision and control: Single vehicle performance. Technical Report AHFD-02-10/MAD-02-1, University of Illinois, Institute of Aviation, Aviation Human Factors Division, 2002.
12. C. D. Wickens, S. Dixon, and D. Chang. Using interference models to predict performance in a multiple-task uav environment-2 uavs. Technical Report AHFD-03-09/MAAD-03-1, University of Illinois, Institute of Aviation, Aviation Human Factors Division, 2003.

8

Multi-UAV Experiments: Application to Forest Fires

J.R. Martínez-de-Dios[1], Luis Merino[2], Aníbal Ollero[1], Luis M. Ribeiro[3], and Xavier Viegas[3]

[1] Robotics, Vision and Control Group, University of Seville, Camino de los Descubrimientos s/n, 41092 Seville (Spain)
{jdedios,aollero}@cartuja.us.es

[2] Pablo de Olavide University, Crta. Utrera km. 1, 41013 Seville (Spain)
lmercab@upo.es

[3] Universidade de Coimbra, Apartado 10131, 3030 Coimbra (Portugal)
luisr1@netcabo.pt,xavier.viegas@dem.uc.pt

Summary. This Chapter presents the application of a multi-UAV system to forest fires. Particularly the experiments carried out with the COMETS system will be presented. After the introduction and motivation, the UAVs, sensors and basic methods are presented. The third section deals with the general description of the fire detection, localization and monitoring. The next sections are devoted to the multi-UAV surveillance and fire alarm detection, fire observation and monitoring, and cooperative fire monitoring. These sections include short summaries of experiments carried out in the Lousã airfield and the Serra de Gestosa, near Coimbra (Portugal).

8.1 Introduction and Motivation

Forest fire is an appropriate scenario for the demonstration of multi-UAV capabilities and performance. Forest fires are highly complex, non-structured environments, where the use of multiple sources of information at different locations is essential. Besides, fire evolution, which is very difficult to be predicted, and the presence of smoke occluding the images requires flexible re-planning and rescheduling of UAVs, which makes this environment a suitable scenario for testing multi-UAV performance.

Forest fires are a serious problem in many countries with high socioeconomic and environmental impacts and also with global consequences on greenhouse effect, desertification and climate change. In the last years, forest fire applications have attracted significant R&D efforts and several new technologies and platforms have been researched and applied including satellite systems [10, 12], conventional piloted aircrafts and ground stations [2, 4, 13].

However, none of these technologies offer a solution to the problem. For instance, satellite-based systems often have low spatial and temporal resolutions. Although they are successfully applied in uniform and low populated areas, they are not appropriate for regions with intensive human activities such as the European Mediterranean basin. Ground stations and conventional piloted aircrafts

A. Ollero and I. Maza (Eds.): Mult. Hetero. Unmanned Aerial Vehi., STAR 37, pp. 207–228, 2007.
springerlink.com

also have drawbacks. The application of more flexible and effective technologies and particularly its usage in operational forest fire fighting conditions is still a strong need.

Besides, forest fire fighting is a dangerous task which originates many casualties every year. The application of UAVs is convenient to reduce the presence of firemen brigades and to provide information to better organize the fire attack.

In [1] the application of a Medium Altitude (27000 feet) Long Endurance (30/40 hours) UAV with 450 lbs payload, was demonstrated in fire experiments. The data received at the ground station were geo-referenced. The process time from data acquisition aboard the UAV, through satellite uplink/downlink, processing into a geo-referenced image data file, to a fire manager in a remote control center was less than an hour.

In the COMETS project, instead of using a single powerful UAV with significant on-board resources but also with high cost, the application of a fleet of lower cost UAVs for forest fire fighting is proposed. These UAVs can be used as local flying sensors providing images and data at short distance.

This Chapter presents experimental results of multi-UAV forest surveillance, forest fire detection, localization and confirmation as well as fire observation and measurement obtained in the COMETS project. Although there is still need of research and development, the results represent a significant step towards the application of UAVs in forest fire operational conditions.

8.2 A Multi-UAV System for Forest Fire Applications

The experiments described in this Chapter were performed in the Serra de Gestosa (see Fig. 8.1) and the airfield of Lousã (Portugal) in May 2004 and 2005. In the region of Centro (Portugal), Lousã is located at one relevant forested areas of Portugal. The experiments were carried out with real fires controlled by firemen in close-to-operational conditions. The ADAI group of the University of Coimbra (Portugal) coordinated the arrangements of fire experiments. Figure 8.2 shows a map and an aerial photograph of the Lousã airfield.

8.2.1 Descriptions of the UAVs

The experiments considered a fleet of three heterogeneous UAVs: the helicopter MARVIN [11], the helicopter Helivision-GRVC [9] and, the airship Karma [5]. These UAVs are presented in Chaps. 5, 6 and 7, respectively. Figure 8.3 shows MARVIN, Helivision-GRVC and Karma UAVs in forest fire experiments at the Lousã airfield.

The heterogeneity of these UAVs is manifold. Complementary platforms are considered: helicopters have high maneuverability and hovering ability, and are suited to agile target tracking tasks and inspection and monitoring tasks that require to maintain a position and to obtain detailed views. Airships have less manoeuvrability but can be used to provide global views or to act as communication relays. Besides, these UAVs are also heterogeneous in terms of on-board

Fig. 8.1. Left: Location of Lousã on general map of Portugal. Top right: Aerial view of Serra de Gestosa (Portugal). The rectangular plots burned in the experiments can be observed in the images. Bottom right: Helivision-GRVC in forest fire experiments at Serra de Gestosa in May 2004.

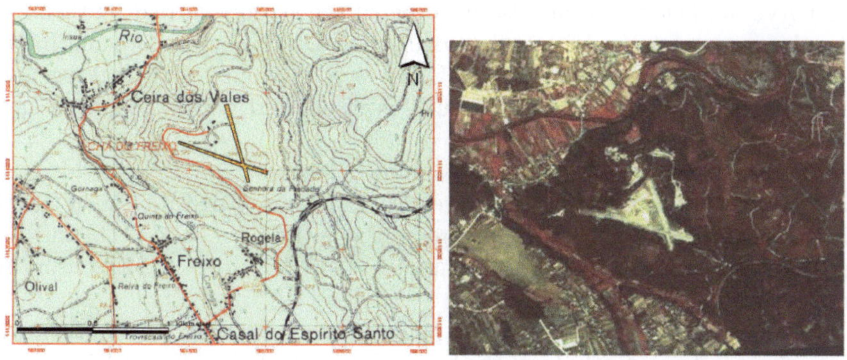

Fig. 8.2. Map and aerial photograph of Lousã airfield

processing capabilities, ranging from fully autonomous aerial systems to conventional radio controlled systems with minimal on-board capabilities required to record and transmit information.

Fig. 8.3. MARVIN, Helivision-GRVC and Karma UAVs in forest fire experiments at the airfield of Lousã (Portugal) in May 2005

8.2.2 Sensors of the Fleet

The UAVs are heterogeneous also in terms of the sensors carried by them. Besides the sensors required for navigation such as DGPS, gyroscopes and Inertial Measurement Units and others, the UAVs were equipped with heterogeneous sensors for environment perception such as visual and infrared cameras and a specialized fire sensor.

Helivision-GRVC is equipped with an infrared camera and a visual video camera. The infrared camera is the low-weight (150 g) un-cooled Raytheon 2000AS camera (see Fig. 8.4 left), which operates in the far infrared band (7–14 μm). The visual camera is a low-weight Camtronics PC-420DPB with 752x582 sensors and lens with focal distance of 6 mm.

Helivision-GRVC has a motorized pan and tilt unit that allows orientating the cameras independently from the body of the vehicle (see Fig. 8.4 right). The unit has encoders to measure the pan and tilt angles.

The infrared and visual cameras of Helivision-GRVC were geometrically calibrated. The GPS, IMU and pan and tilt encoders of the camera positioning system allow obtaining the camera position and heading. These data are used to geolocate objects on the image plane by projecting them over a known elevation map as described in Chap. 4.

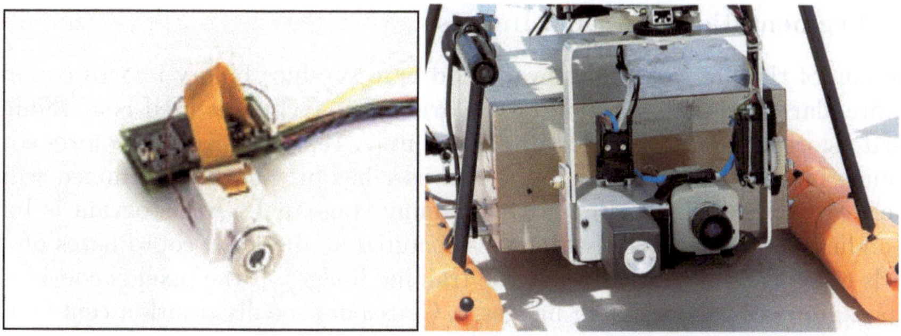

Fig. 8.4. Left: detail of the Raytheon 2000AS OEM infrared camera. Right: Infrared and visual cameras mounted on Helivision-GRVC pan and tilt unit.

Fig. 8.5. Fire sensor (left) and digital photo camera (right) on-board MARVIN

MARVIN carries a Hamamatsu UV-Tron fire detector, whose main component is a photodiode set-up to limit its sensibility to the band of [185, 260] nm, normally associated to fires radiation. The output of the sensor is a scalar value, proportional to the radiation energy, received every two seconds. Being a magnitude sensor, it is not possible to determine if a measure is due to a big fire far away or to a nearby small fire. Also, the sensor cannot directly provide the position of the fire. MARVIN also carries a Canon Powershot S45 digital photo camera. Figure 8.5 shows the fire sensor and digital photo camera mounted on MARVIN.

Karma carries a stereo bench with two visual digital IEEE1394 colour cameras (1024x768 pixels). The stereo system is mainly applied to obtain 2D and 3D terrain maps.

8.2.3 Fire Segmentation

Forest fire perception requires algorithms capable of identifying the fire in images from the infrared and visual images and in the data generated by the fire sensor. This Section describes these segmentation techniques.

Fire Segmentation in Visual Images

The aim of the fire segmentation method is to produce binary images containing fire alarms while discarding false alarms. The technique used is a training-based algorithm similar to those described in [7, 15]. The method requires some training images in which an experienced user has previously determined which pixels correspond to the fire. In the training stage, a RGB histogram is built by adding Gaussian-type distributions centered at the RGB coordinates of the pixels considered as a fire pixel in the training images. If the pixel is considered as background in the training images, a Gaussian-type distribution centered at the RGB coordinates is subtracted from the RGB histogram.

Finally, this RGB histogram is thresholded and a look-up table for the RGB colour space is built. The look-up table contains a Boolean value indicating whether the colour represents fire or background. In the application stage, the RGB coordinates of the pixels are mapped in the trained look-up table and are considered fire pixels if the value in the look-up table is '1' and, background otherwise. Figure 8.6 shows one visual image of a Gestosa experiment and the corresponding segmented image.

Fig. 8.6. Left: original colour visual image from a Gestosa fire experiment. Right: corresponding segmented image.

Fire Segmentation in Infrared Images

The aim of the processing of infrared images is to produce binary images containing fire alarms while discarding false alarms. Since fires appear in infrared images as high intensity regions, the first step in fire segmentation is to apply a threshold value. The following step is to apply heuristic rules to discriminate false alarms.

The infrared camera used in the experiments was a low-cost OEM non-thermal camera. It does not provide temperature measures but estimations of the radiation intensity throughout the scene. Black and white colours represent low and high radiation intensities, respectively. For robust fire segmentation, the thresholding technique should consider the particularities of the application. For instance, in the current state of technology, the detectors of miniaturized infrared cameras still have low sensitivity and require high detector exposure periods to generate the images. Thus, the high frequency vibrations induced by the UAV engine often originate blurs in the infrared images.

The solution adopted was to use the training-based thresholding method described in [3]. Its main idea is to extract the particularities of a computer vision application and use them to supervise a multiresolution histogram analysis. The technique is applied in two stages: training and application (see Fig. 8.7). The training stage requires a set of training images and their corresponding desired threshold values given by an experienced user. The training stage identifies the conditions under which pixels should be considered to belong to the object of interest. These particularities are introduced in a system via the ANFIS training method [6]. In the application stage, features of the image are used to determine a suitable threshold value according to these particularities. A detailed description can be found in [3].

The adaptation of the method to this application was carried out with training infrared images with different illumination conditions, different image backgrounds and different objects including fires and false alarms (i.e. heat emitting sources such as car engines).

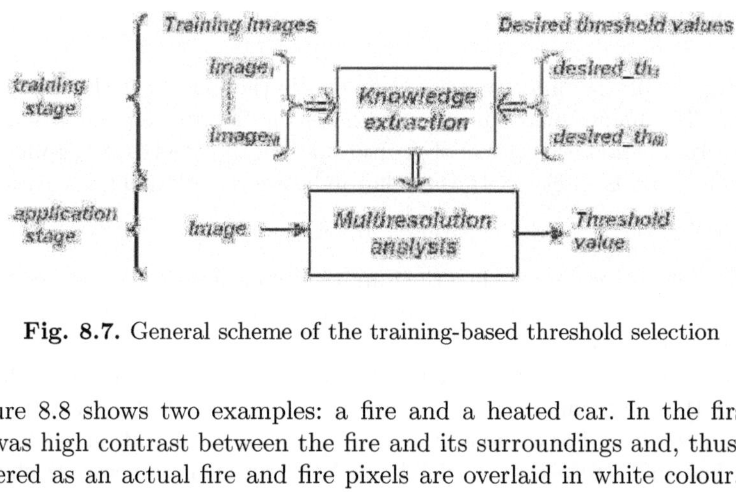

Fig. 8.7. General scheme of the training-based threshold selection

Figure 8.8 shows two examples: a fire and a heated car. In the first case, there was high contrast between the fire and its surroundings and, thus, it was considered as an actual fire and fire pixels are overlaid in white colour. In the second case, the contrast was low and this early processing discarded the alarm as false.

Fusion of Infrared and Visual Images for Fire Segmentation

Helivision-GRVC UAV carries both, an infrared and a visual camera (see Fig. 8.4). These sensors can be combined obtaining what it can be considered a "multispectral" camera, which can provide more robust results for fire detection purposes. This Section describes how this can be accomplished.

Assume that both cameras share the centre of projection. Figure 8.9 presents a scheme of the geometry of this configuration.

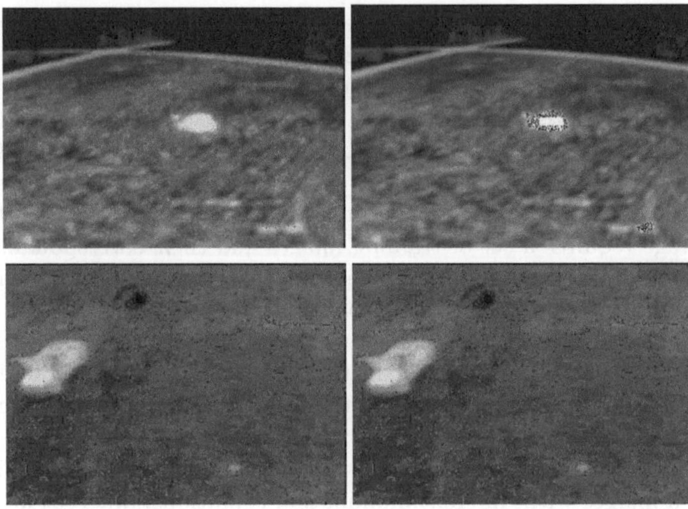

Fig. 8.8. Two infrared images from a fire (top) and from a heated car engine (bottom). Top: there is high contrast between the fire and its surroundings and the alarm is considered as fire. Bottom: the contrast is low, the alarm is discarded as false and no object is segmented.

Let $\mathbf{m}_{\text{IR}} = [u\,v\,1]^T$ and $\mathbf{m}_{\text{VIS}} = [u'\,v'\,1]^T$ be the images at the same instant of a point $\mathbf{X}$ in homogeneous pixel coordinates on the infrared and visual images respectively. Then, if the centres of projection of both cameras are assumed to be coincident (point $\mathbf{C}$ in Fig. 8.9), the relation between both images is given by:

$$s\mathbf{m}_{\text{IR}} = H_\infty \mathbf{m}_{\text{VIS}} \tag{8.1}$$

where H_∞ is a 3×3 matrix called the infinity homography and s is a scale factor. H_∞ can be calculated knowing at least four correspondences among points or lines in both images.

There are many algorithms for point matching between images of the same modality. However, this is a challenging problem when dealing with images of different modalities. For the experiments in Lousã, the calibration has been done using a known pattern that is visible on both types of cameras. It should be noticed that H_∞ is computed only once provided that the relative orientation of the cameras and their internal calibration does not change.

In the system considered, the centres of projection of both cameras will not be actually coincident, but the equations above hold if the distance between the centres of projection is small compared with the distances of the points of the scene with respect to the cameras. Figure 8.10 shows some results of the combination of infrared and visual images.

Using the technique presented above, any pixel in the infrared camera can be related to a corresponding pixel in the visual camera. This will be used to improve the results on fire detection, combining the results from the algorithms for fire segmentation in infrared and visual images.

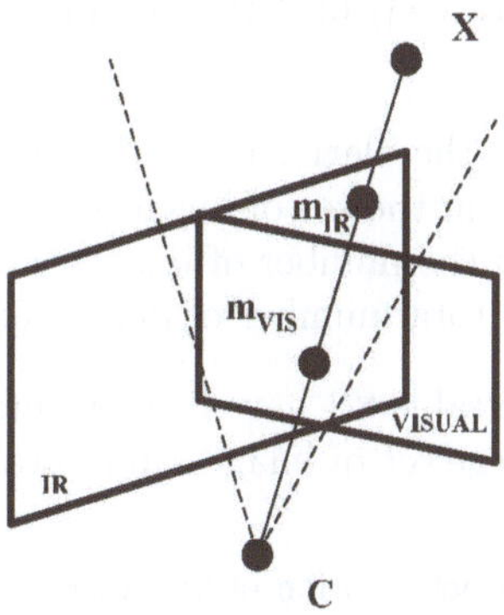

Fig. 8.9. Geometry of the configuration of the cameras

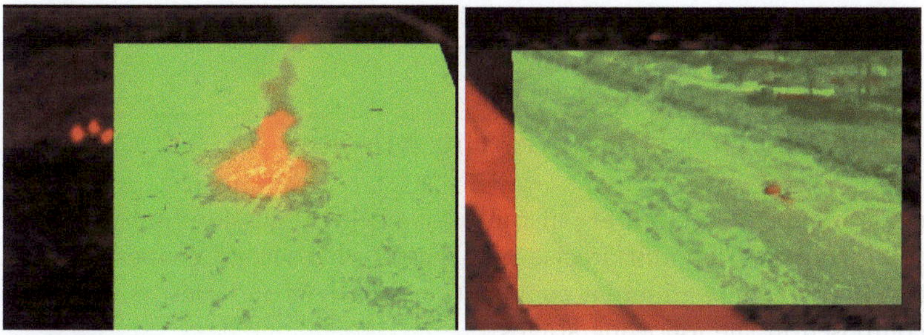

Fig. 8.10. Combination results. The infrared and visual images are presented together as the red and green fields of a colour image.

Fire Identification in Data from the Fire Detector

The fire sensor provides a scalar value indicating the presence of fire. This value is proportional to the amount of radiation received in a narrow band adjusted to the emissions of fire. Thus, fire is segmented by applying a threshold, so that a Boolean value is obtained, indicating that a fire is present in the field of view of the sensor.

The threshold can be adapted depending on the application. A lower threshold increases the probability of detection (which can be convenient during a detection stage), but on the other hand produces more false alarms. A higher threshold rejects more false alarms while reducing the detection capabilities (interesting for confirmation stages).

Characterization of the Fire Segmentation and Identification

The cooperative perception techniques described in Chap. 4 requires a probabilistic characterization of the fire segmentation algorithms. The algorithms are modelled by the probabilities P_D of detection and P_F of false positive outputs defined as follows:

- P_D is the ratio between the alarms correctly detected and the total number of fire alarms presented in the set of images.
- P_F is the ratio between the number of images where the algorithm detected fire incorrectly and the total number of images of the sequence.

The values shown in the table 8.1 have been experimentally computed for the three algorithms with a large set of images and data, including real fires.

Table 8.1. Characterization of fire segmentation algorithms

	Infrared camera	Visual camera	Fire detector
P_D	100%	89,2%	95%
P_F	8,9%	3,1%	1%

8.3 General Description of the Mission

The experiments carried out are instances of a general mission that performs fire detection, confirmation and precise localization with several cooperating UAVs, and that could include different combinations of UAVs and sensors. The mission is decomposed in the following stages:

- *alarm detection*,
- *alarm confirmation*,
- *fire observation and measurement.*

The main idea is to simulate the performance of forest fire fighting protocols currently applied in many countries. In fact, these three stages are common in current operational forest fire fighting operations. Forest surveillance is often carried out by experienced operators in watch towers, by aerial patrolling or by automatic fire detection systems. Detected alarms are confirmed by using additional information, which often involves that a piloted helicopter is sent for visual confirmation. If the alarm is confirmed, the fire extinguishing protocols are started. These protocols require to know the state of the forest fire including its location, its severity and several fire geometry features, which are currently visually estimated by expert firemen.

The *alarm detection* stage starts by searching potential fires. In this stage, the overall area to be surveyed by the fleet is divided in searching regions for MARVIN and Helivision-GRVC. For fire search MARVIN uses its fire sensor and Helivision-GRVC uses its infrared camera. Karma is sent to obtain a global view of the overall area. At this location Karma acts as communication relay. When any of the UAVs detects a potential fire alarm, this stage concludes and the *alarm confirmation* stage starts.

In the *alarm confirmation* stage the tasks for MARVIN and Helivision-GRVC are replanned. The UAV that detected the alarm is commanded to hover at a safety distance from the fire alarm. The other UAV is sent to the alarm location in order to confirm the alarm by using its own sensors. The data from different sensors in different UAVs are merged to confirm or discard the alarm. If the alarm is found to be false, then the *alarm confirmation* stage finishes and the *alarm detection* stage is resumed. If the alarm is confirmed as a fire, then the *fire observation* stage starts.

In the *fire observation* stage the tasks for MARVIN and Helivision-GRVC are re-planned: both are commanded to hover on to the fire alarm and to obtain stabilized images and data of the fire from different views.

Figure 8.11 shows two images taken from forest fire multi-UAV experiments. In the following sections, these stages are described with more details and illustrated with results from field experiments.

Fig. 8.11. Left: Karma and Helivision-GRVC involved in a fire experiment at the Lousã airfield (Portugal). Right: a view of the airfield from MARVIN.

8.4 Multi-UAV Surveillance and Fire Alarm Detection

This stage can be decomposed in two essential steps. In the first step, the global region to be surveyed is divided in regions for each UAV. For such a division the Control Centre takes into account the capabilities of each UAV including the flight speed and altitude of each UAV, as well as the sensing width which depends of the field of view of the cameras or detectors on-board each UAV. The UAVs cover their convex surveillance regions by describing back and forward rectilinear sweeps. The division of the overall region is carried out by minimizing the number of sweep turns, which involve significant time for the UAVs to stop, rotate and then accelerate for the following sweep (more details on can be found in [8]). Once the Control Centre has divided the global area among the UAVs, it assigns the waypoints required for each UAV to describe the assigned trajectory.

In the second step, each UAV of the fleet follows the waypoints commanded by the Control Centre and performs individual fire alarm search by applying the fire segmentation algorithms described in Sect. 8.2.3.

The data from the MARVIN fire sensor are processed to evolve a fire probability grid over the searching zone by using the techniques described in Chap. 4. The grid covers 310x400 square meters and each cell corresponds to an area of 1 m^2. Initially all the cells are set to a probability value of 0.5. Figure 8.12 shows the initial stages of the evolution of the probability values of the grid every 40 seconds. Each triangle represents the area covered by the fire sensor (several cells of the grid) at different time instants and, its colour represents the probability of having a fire alarm in this grid cell. Black colour is low probability and white is high probability. Connected cells with high probability (higher than 0.7) are considered as fire alarms and their locations are obtained.

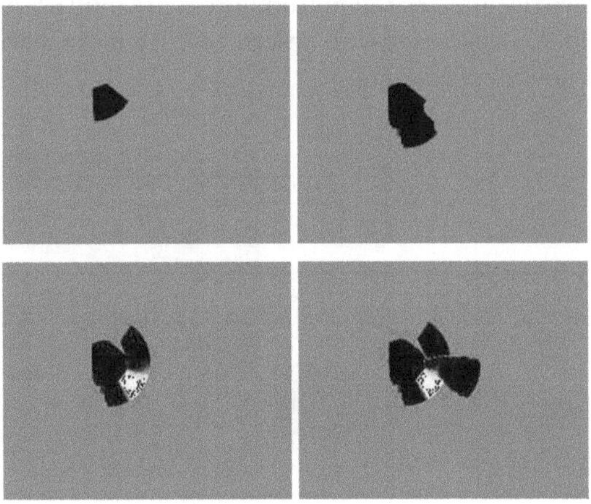

Fig. 8.12. Evolution of fire probability of the grids cells. Each triangle represents a cell of the grid. White colour indicates grid cells with high fire probability and black colour, with low probability.

Figure 8.13 shows a georeferenced schematic map of the Lousã airfield. The solid polygonal object represents the concrete area where the UAVs take off and land. The axes are shifted UTM coordinates in meters. In the whole duration of this stage, three potential alarms are detected with the MARVIN fire sensor. Only one of the three fire alarms is a true fire. The position of the actual controlled fire is marked with a solid square.

Also, Helivision-GRVC applies the fire segmentation algorithm in infrared images described in Sect. 8.2.3. If a fire alarm is segmented, the fire segmentation in visual images is applied over the Helivision-GRVC visual image in order to confirm the alarm. If the same alarm is segmented in both images, then Helivision-GRVC sends the alarm to the Control Center.

Figure 8.14 top shows the original and segmented infrared images of a fire alarm detected in a field experiment in May 2004. Figure 8.14 centre shows the corresponding original and segmented visual images. Figure 8.14 bottom displays the overlapping between both segmented images.

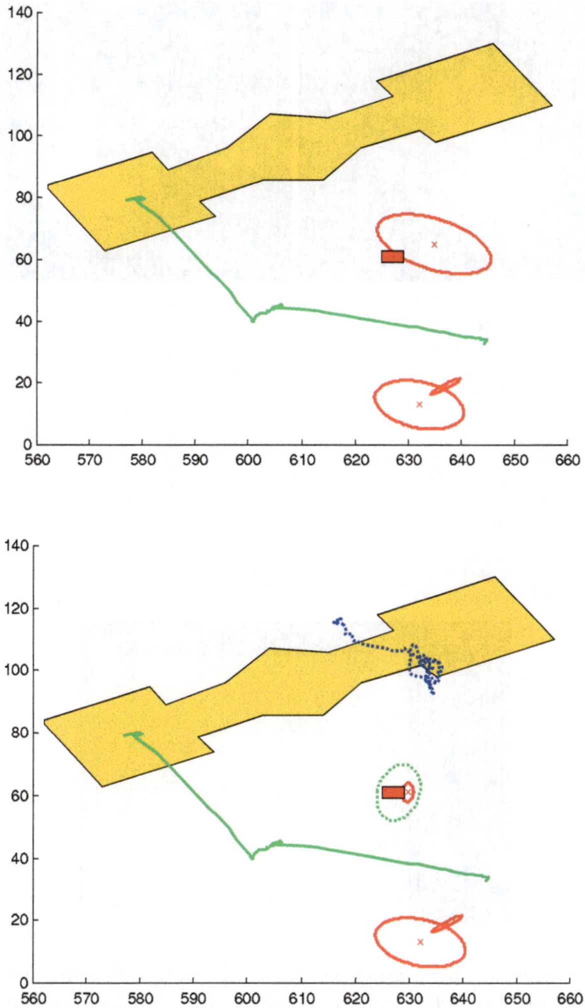

Fig. 8.13. Top: alarm detection stage. Fire alarms detected using the fire sensor. The ellipses represent the uncertainties in the computed positions of the alarms. The square indicates the actual position of the fire. The trajectory of MARVIN is shown. Bottom: alarm confirmation stage. New measures from Helivision-GRVC (dotted ellipse) are used to refine the location of the alarms. Helivision-GRVC trajectory is also shown (dotted).

When any of the UAVs of the fleet detects a fire, it georeferences the alarm by using the techniques described in Chap. 4 and sends the georeferenced location of the alarm to the Control Centre. Then, the *alarm detection* stage concludes and *alarm confirmation* starts.

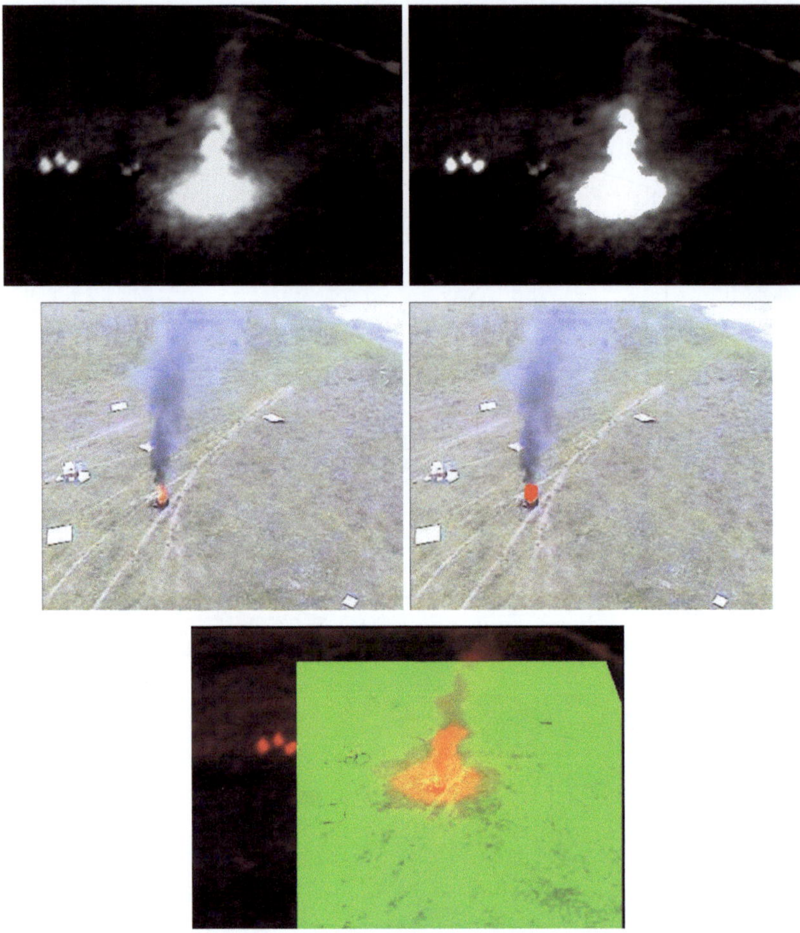

Fig. 8.14. Top: original and segmented infrared images of a fire alarm detected in a field experiment in May 2004. Centre: corresponding original and segmented visual images. Bottom: overlapping between segmented infrared (red field of the colour image) and visual images (green field of the colour image).

8.5 Cooperative Alarm Confirmation

In the *alarm confirmation* stage, the missions of the UAVs are re-planned:

- the UAV that detected the alarm is commanded to hover at the location of the alarm,
- the remaining UAVs of the fleet are commanded to go to the location of the alarm.

When the UAVs are at the location of the alarm, the images and data from all the UAVs are processed. The results of the fire segmentation algorithms are

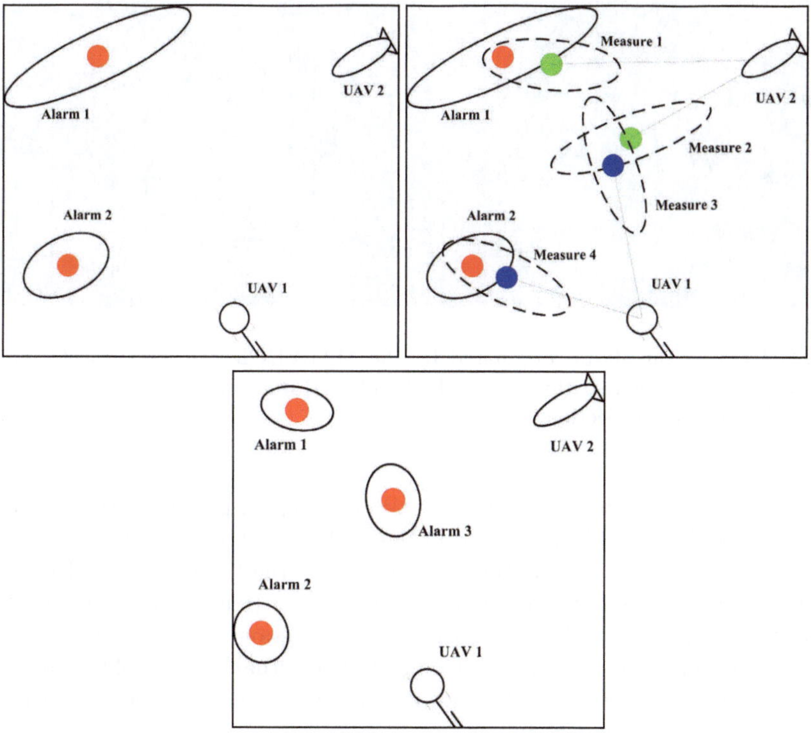

Fig. 8.15. Top left: prediction. The previous detected alarms and their uncertainties (presented as ellipses). Top right: capture of new measures. These measures are associated to the tracks of the currently detected alarms. Bottom: the update stage reduces the uncertainties of the tracks with the new data inputs. New tracks are added.

merged by using the cooperative detection technique described in Chap. 4 to confirm or discard the alarm. Figure 8.15 illustrates the scheme of cooperative alarm confirmation.

Figure 8.16 shows the fire alarms from the MARVIN fire sensor projected onto the segmented infrared images of Helivision-GRVC. The ellipses represent the uncertainty on the locations of the alarms of the fire sensor and the white patch is a region segmented as fire in the infrared image. In this example, one of the three uncertainty ellipses intersects the area of the alarm generated by the infrared images. The probability of this MARVIN alarm is enforced by the alarm of Helivision-GRVC and its probability is increased accordingly. The data of these associated alarms are used to update the probability of being fire for each alarm and also to refine the estimation of its location. The other two alarms originated by the MARVIN fire sensor are not associated to the alarm originated by the Helivision-GRVC infrared camera, and, thus, their probabilities alarms are decreased.

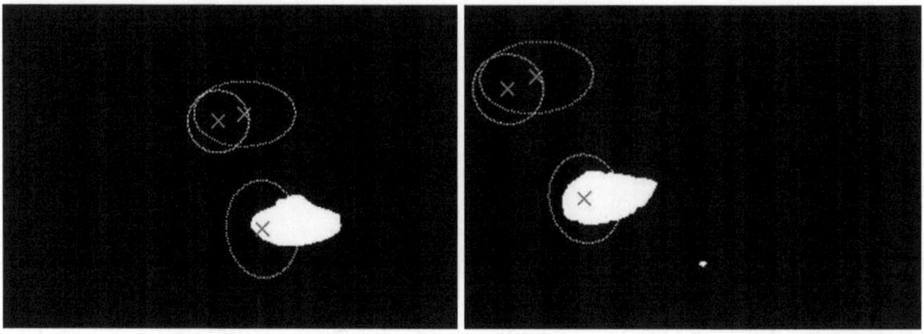

Fig. 8.16. Current alarms originated by the MARVIN fire sensor projected on the image plane of segmented Helivision-GRVC infrared images. The ellipses indicate the uncertainty on the projection.

Figure 8.17 shows how the uncertainties in the position of the true alarm are recursively reduced, while the probabilities of the false alarms drop to values close to 0, when the alarm information obtained from the fire sensor of MARVIN is combined with the data from the infrared camera of Helivision-GRVC. Table 8.2 presents the position of the fire alarm (mean and standard deviation) estimated with the fire sensor and the infrared camera. The actual location of the controlled fire measured using a GPS is also shown. The similarities between the actual and estimated position are evident.

Table 8.2. Estimated and true location of the fire and uncertainties

	Easting	Northing	Height
True location of the fire	564627	4443961	200
Final estimated location (fusion)	564628.9	4443961.4	200.04
Estimated standard deviation	1.5	2	0.28

When the alarm is confirmed as actual fire or discarded as a false alarm, the *alarm confirmation* stage finishes and the *fire observation* stage starts.

8.6 Fire Observation and Monitoring Using an UAV

Fire monitoring, in the field of forest fire fighting, could be defined as the computation in real-time of the dynamic evolution of forest fire parameters, such as the shape and location of the fire front, the maximum height of the flames and others [14]. This has been done traditionally by visual inspection carried out by experts. Also, photogrametric techniques have been applied for analysis after a fire has taken place.

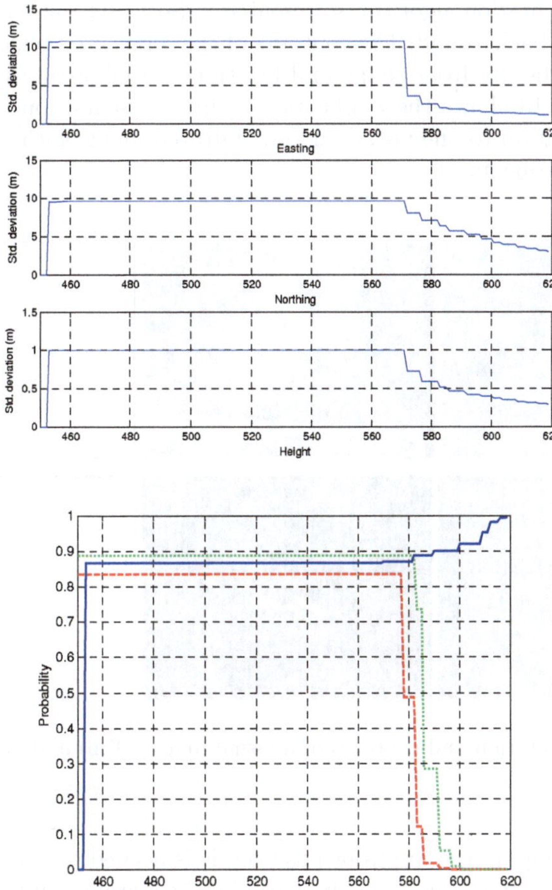

Fig. 8.17. Evolution in the localization of the alarm. Top: The graphic shows the estimated standard deviation for the location of the alarm (Easting, Northing and Height). The alarm is obtained from the fire sensor data at time 450. The initial errors are high. Around time 570, images from Helivision-GRVC are used to confirm the alarm and to refine the position. Bottom: evolution of the probabilities of the three alarms.

The development of an automatic system based on vision for online fire monitoring would be very helpful. However, monitoring real fires poses some strong problems, such as the possibility of placing sensors in the correct locations for fire observation. Usually, conventional piloted helicopters are used to approach the fire in fire fighting activities. However, these operations involve high risks. One solution would be to use cameras on board high-manoeuvrability UAVs.

In this section, a procedure for the estimation of the fire front evolution from images gathered by one UAV is briefly presented.

The fire segmentation algorithms presented in Sect. 8.2.3 are used to separate the fire from the background. The contours of the segmented regions contain the information on the fire front shape and location as well as on the height of the flames. Figure 8.18 shows the application of fire segmentation algorithms and contour computation to one visual and one infrared image gathered by an UAV in fire field experiments.

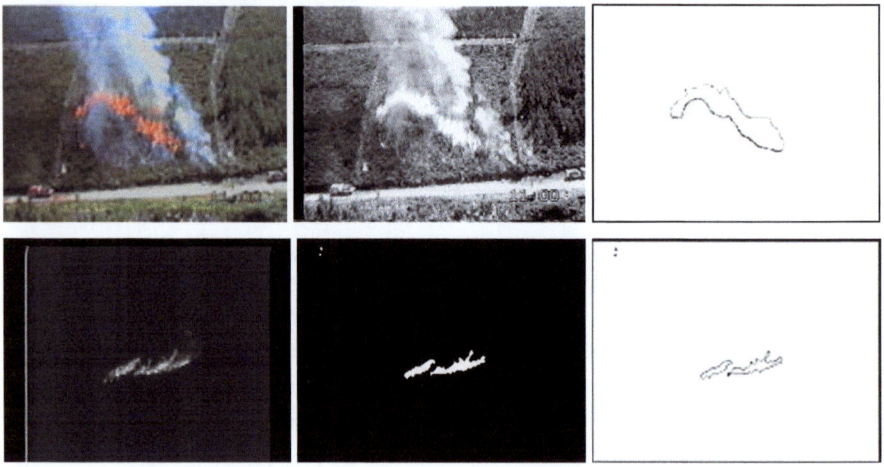

Fig. 8.18. Segmentation and contours of a visual image (Top) and an infrared image (Bottom)

In order to compute the fire front position, it is needed to distinguish, among the pixels of these contours, the subset of pixels corresponding to the fire base contour and the corresponding pixels of the top of the flames.

The dynamic characteristics of the fire are used to characterize the fire contours. In general, the pixels corresponding to the top of the flames flicker, while the pixels corresponding to the fire base move slowly. Thus, in order to determine which pixels correspond to the fire base contour and which to the flames contour, a temporal low pass filter is applied over the fire contour images.

In order to analyze these dynamic characteristics, we need to remove previously the background motion. This is accomplished by using the image stabilization procedure presented on Chap. 4. Figure 8.19 shows three images at different time instants of a forest fire experiment from a helicopter. The images have been stabilized, and the fire contour is overlaid, distinguishing between the fire front and the pixels of the top of the flames.

The computation of the fire front position is done obtaining the 3D position of the pixels belonging to the fire fronts. The geolocation procedure of Chap. 4 is used for this purpose. Figure 8.20 shows the estimated fire front position every 10 seconds for the images of the sequence of Figure 8.19. The gaps observed are due to smoke occlusions.

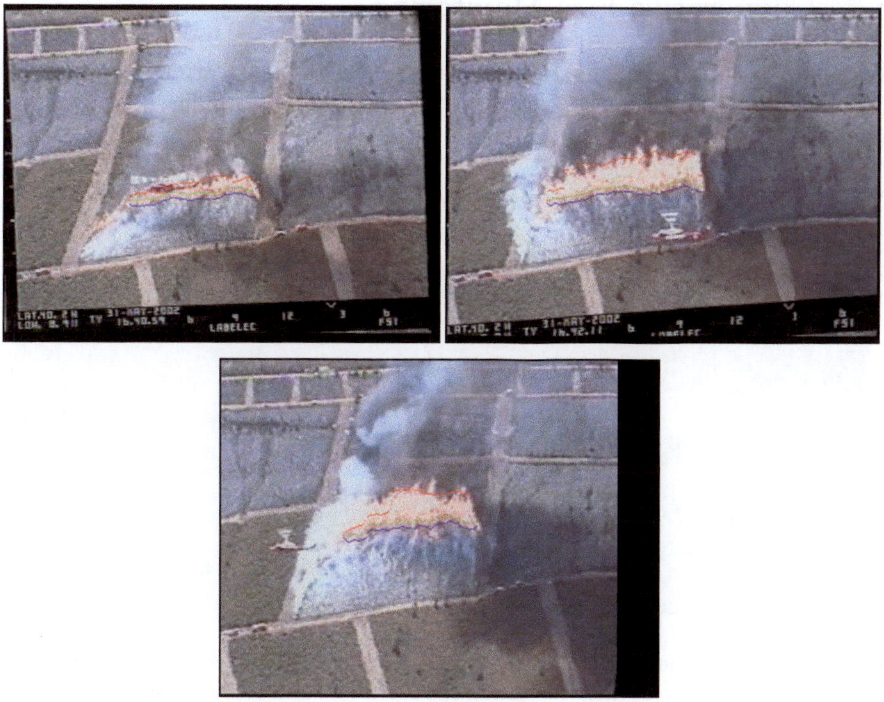

Fig. 8.19. Three images from a helicopter of a forest fire experiment

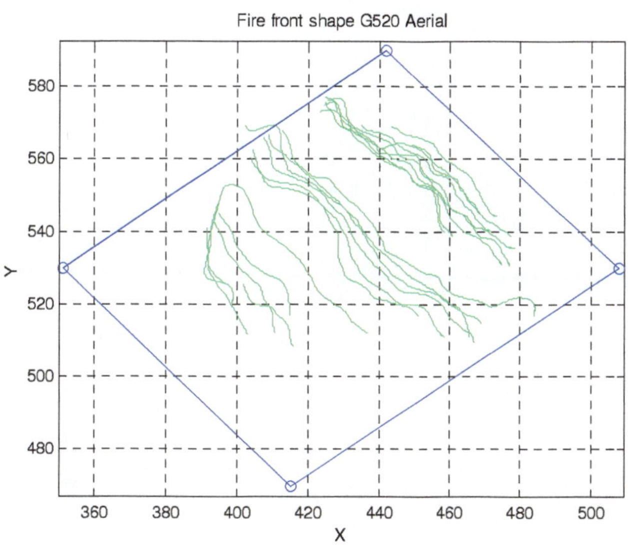

Fig. 8.20. Fire front location for the fire experiment of the images shown in Fig. 8.19. The location of the fire front is estimated every 30 seconds.

8.7 Cooperative Fire Monitoring

In this stage MARVIN and Helivision-GRVC keep on hovering on to the fire (approximately forming 120° with the fire) and sending stabilized sequences of images of the event (an operator could then observe the dynamic evolution of the fire). Figure 8.21 shows two visual images taken by Helivision-GRVC and MARVIN obtained in the *fire observation* stage.

Fig. 8.21. During the *fire observation* stage, sequences of stabilized images from different points of view are obtained by using the MARVIN and Helivision-GRVC cameras

Fig. 8.22. Fire observation by using the MARVIN and Helivision-GRVC visual cameras. The fire front shape is marked on the images.

The architecture described in Chap. 2 allows to synchronize the vehicles to send images close in time. The images are previously stabilized in real-time, using the procedures described in Chap. 4. With these stabilized images, and using the fire segmentation algorithms, it is possible to determine parameters of the fire evolution. Figure 8.22 shows images from MARVIN and Helivision-GRVC of the same fire, with the fire front marked.

8.8 Conclusions

Relevance to applications is a main issue in multi-robot systems and particularly in multi-UAV systems. This Chapter has presented the application of the COMETS multi-UAV system in a practical application with significant interest: forest fires. As far as we know this is the first experimental demonstration of the interest of a multi-UAV system in a mission consisting in fire detection, confirmation, localization and monitoring.

These experiments and demonstrations have been carried out in central Portugal, near the city of Coimbra in May 2003, 2004 and 2005. They involved the MARVIN autonomous helicopter (Chap. 5), the Karma airship (Chap. 7), and the Helivision-GRVC teleoperated helicopter (Chap. 6). The decisional architecture presented in Chap. 2, the communication system from Chap. 3 and the perception techniques described in Chap. 4 have been also applied.

The obtained results confirm the interest of the UAV technology and presented methods. It also points to the application of some of the technologies and methods presented in this book in operational forest fire conditions. These operational conditions require the use of different operational UAV platforms, including the integration in the multi-UAV system of fixed wing aircrafts and, in general, the use of UAVs platforms having more flight range, endurance and adaptation to typical forest fire scenarios with important wind velocity.

References

1. V.G. Ambrosia. Remotely piloted vehicles as fire imaging platforms: The future is here! *Wildfire Magazine*, May–June 2002.
2. B.C. Arrue, A. Ollero, and J.R. Martinez de Dios. An intelligent system for false alarm reduction in infrared forest-fire detection. *IEEE Intelligent Systems*, 15(3):64–73, 2000.
3. J.R. Martinez de Dios and A. Ollero. A multiresolution threshold selection method based on training. volume 3211 of *Lecture Notes in Computer Science*, pages 90–97. Springer, 2004.
4. D. Dierre, H. Hoff, and M. Bouchet. RAPSODI, rapid smoke detec-tion and forest fire control. In *Proceedings of the Research Special Session of the Euromediterranean Wildfire Meeting*, pages 89–96, Hyères, France, October 2000.
5. E. Hygounenc, Il-Kyun Jung, P. Souères, and S. Lacroix. The Autonomous Blimp Project of LAAS-CNRS: Achievements in Flight Control and Terrain Mapping. *The International Journal of Robotics Research*, 23(4–5):473–511, 2004.

6. Jyh-Sing Roger Jang. ANFIS: Adaptive-network-based fuzzy inference system. *IEEE Transactions on Systems, Man, and Cybernetics*, 23:665–684, 1993.
7. R. Kjedlsen and J. Kender. Finding skin in colour images. In *Proceedings of the International Workshop on Face and Gesture Recognition*, pages 144–150, 1996.
8. I. Maza and A. Ollero. Multiple UAV cooperative searching operation using polygon area decomposition and efficient coverage algorithms. In *Proceedings of the 7th International Symposium on Distributed Autonomous Robotic Systems (DARS)*, pages 211–220, Toulouse, France, June 2004.
9. A. Ollero, J. Alcázar, F. Cuesta, F. López-Pichaco, and C. Nogales. Helicopter teleoperation for aerial monitoring in the COMETS multi-UAV system. In *Proceedings of the 3rd IARP Workshop on Service, Assistive and Personal Robots (IARP)*, pages 137–142, Madrid, Spain, October 2003.
10. Y. Rauste. Forest fire detection with satellites for forest fire control. In *Proceedings of the XVIII Congress of the International Society for Photogrammetry and Remote Sensing*, volume XXXI, pages 584–588, Vienna, Austria, July 1996.
11. V. Remuß, M. Musial, and G. Hommel. MARVIN – an autonomous flying robot-based on mass market components. In *Proceedings of the Aerial Robotics Workshop of the IEEE/RSJ International Conference on Intelligent Robots and Systems (IROS)*, pages 23–28, Lausanne, Switzerland, September–October 2002.
12. I.F. Tourné, J. Gonzalo, and P. Yagüe. The fuego system. In *Proceedings of the Research Special Session of the Euromediterranean Wildfire Meeting*, pages 118–122, Hyères, France, October 2000.
13. A.B. Utkin, A. Fernandes, F. Simoes, R. Vilar, and A. Lavrov. Forest fire detection by means of lidar. In *Proceedings of IV International Conference on Forest Fire Research / 2002 Wildland Fire Safety Summit*, pages 18–23, Coimbra, Portugal, November 2002.
14. D.X. Viegas. Innovations and solutions in fire behavior prediction issues. forest fires: needs and innovations. In *Proceedings of the Delfi International Symposium*, pages 164–173, Athens, Greece, 1999.
15. III Walter Phillips, Mubarak Shah, and Niels da Vitoria Lobo. Flame recognition in video. *Pattern Recogn. Lett.*, 23(1–3):319–327, 2002.

9

Conclusions and Future Directions

Aníbal Ollero and Iván Maza

Robotics, Vision and Control Group, University of Seville, Camino de los Descubrimientos s/n, 41092 Seville (Spain)
{aollero,imaza}@cartuja.us.es

Unmanned aerial vehicles offer advantages for many applications when comparing with their manned counter parts. They preserve human pilots of flying in dangerous conditions that can be encountered not only in military applications but also in other scenarios involving operation in bad weather conditions, or near to buildings, trees, civil infrastructures and other obstacles.

Furthermore, there are commercial applications, such as inspection of infrastructures or electrical lines, in which the use of low cost UAVs can produce significant cost savings when comparing to conventional aircrafts. Moreover, longer endurance of HALE and MALE platforms could provide benefits in applications such as environmental monitoring, communications and others. Then, the prospects for market growth of UAV-based applications are very good and it is expected that in the next 20–30 years, cost-effective UAVs will substitute manned aircraft in many missions and will open new application markets.

This book has been devoted to multi-UAV systems. The benefits of the multi-UAV approach when comparing to the use of a single UAV can be summarized as follows:

- Increased coverage in surveillance by minimizing delays in the event detection.
- Decreased time in exploration, mapping and other missions.
- Improved reliability by avoiding dependencies of a single UAV.
- Decreased uncertainties by providing simultaneous information and measurements from different observation points.
- Makes possible the application of teaming involving multiple aircrafts with different and complementary characteristics and sensors.

The characteristics of the UAV platforms impact the performance of the multi-UAV team. Thus, the following developments will facilitate the penetration of UAV technology in the civil market:

New platforms: Flight endurance and range of the currently available low cost unmanned aerial vehicles, such as the helicopters and airships presented in Chaps. 5, 7 and 6 of this book, are very limited. New platforms are required for many applications. This includes large scale aircraft-like systems for long

A. Ollero and I. Maza (Eds.): Mult. Hetero. Unmanned Aerial Vehi., STAR 37, pp. 229–232, 2007.
springerlink.com

range missions, medium size platforms, mini UAVs and micro UAVs and very small scale systems (few centimeters) for indoor operations.

Autonomy: The application of new methods and technologies in avionics and robotic systems is required in future UAVs to minimize the activity of the ground operators. Chapters 5 and 6 have presented control systems for helicopters and airships. Autonomous take-off and landing have been demonstrated in the work described in Chap. 5. However, the landing in unknown terrains and mobile platforms still requires significant efforts. The same applies for the implementation of autonomous robotic functions such as obstacle avoidance and autonomous tracking. These autonomous functions require suitable environment perception functions and robotic architectures. In spite of some recent developments and demonstrations more efforts are still required for the efficient reliable implementation of these functionalities in commercial systems.

Ground control station and operator interfaces: It involves the adoption of interfaces and systems to monitor the activity of UAVs and to facilitate the intervention of the operators when needed. The application of telerobotics concepts and new multimedia interface technologies, as it is mentioned in Chap. 7 of this book, will favor new implementations. Easy transportation and deployment is also a need for many applications.

Reliability: It is related to the platform itself (mechanics, power system, electronics) and to the implementation of the above mentioned autonomous functions in a variety of different conditions. High dependability properties will be essential in many missions, particularly when considering activities in populated areas. Then, the design and implementation of fault detection and identification techniques as well as new fault tolerant control methods in unmanned aerial vehicles is required.

Application sensors: Development and integration of best adapted on-board sensors, ranging from low-cost, low energy and low accuracy sensors, to high precision sensors for the long distance detection and the close monitoring of events. In some applications new efficient sensor data fusion techniques could improve the performance of individual sensors, as it is shown in Chap. 8 for the fire detection and monitoring.

Affordability: Development of cost-effective platforms, with adequate payload capacity, well-suited to specific application needs and well-defined missions. This is also related to the development of modular platforms and interoperable systems that could be used with a variety of payloads, missions and applications. Usability and flexibility will also be required to adapt the UAVs to their missions and context of operations.

The following paragraphs are specifically devoted to the coordination and cooperation of multiple unmanned aerial vehicles.

As far as the cooperation of unmanned aerial vehicles is concerned, more research and development activities are required to implement and demonstrate the higher cooperation levels in the architecture presented in Chap. 2. Particularly, the implementation of fully decentralized architectures could provide benefits when considering scalability and reliability issues. This implementation requires the above mentioned increase in UAV autonomy, which could be achieved by adopting the progress in embedded computing systems and new miniaturised sensors. Furthermore, new cooperation strategies that consider explicitly fault tolerance and reliability issues are required.

Another new trend could be the development of new control and cooperation techniques for tasks requiring strong interactions between vehicles and between vehicles and the environment, such as lifting and transporting by means of the cooperation of several helicopters carrying the same load. This can be justified due to the strong payload limitation of individual low cost UAVs, and the very high cost of helicopters with important payload capacity. However, this technology has been only tested in simulation and only recent projects dealing with the demonstration with real UAVs have been launched.

The communication technologies will obviously play an important role in the cooperation of multiple unmanned aerial systems. Chapter 3 of this book presents a communication system designed, tested and implemented during the COMETS project. This system can be applied with different communication technologies. The evolution of these technologies could lead to different implementations. This includes the use of more reliable communication and high-bandwidth links integrating the UAV with the ground infrastructure and with other UAVs and unmanned systems in general. This is related to both the individual UAV communication system and the networking technologies by the adoption of new technologies for mobile systems, such as the WiMAX.

In general, the development of the ground infrastructure deserves special attention. Usually this infrastructure is not available, or the communication range required for the UAV operation is too large for the existing technology. In general, the market application of UAVs requires not only low cost vehicles but also infrastructure, platforms and systems to integrate them. Thus, for example, the integration with ground sensor and communication networks could have an important impact in the development of new products. This approach requires the development of new cooperative fault-tolerant aerial/ground-based perception techniques for object and event recognition providing reactivity to changes in the environment. This represents a natural extension of the cooperative perception tools based on aerial sensing presented in this book.

Another related field is the integration of UAVs with the new technologies in Wireless Sensor Networks (WSNs). In fact, a UAV can act as a "data mule" or relay node of the wireless sensor network providing or repairing the connectivity of the network. Even more, a team of UAVs can be used for the deployment or repairing of the network by transporting and locating wireless sensor nodes or communication equipment. Recently, new research projects dealing with these advanced concepts have been launched.

The ground infrastructure also includes the development of ground stations for the monitoring and teleoperation of multiple unmanned aerial vehicles. Some of the concepts in Chap. 7 can be applied, but significant research and development efforts are needed to produce a ground station in which a minimal number of human operators can monitor and teleoperate a team of UAVs.

Moreover, the practical application of a team of aerial vehicles will require the integration with piloted aerial vehicles. Thus, for example, this is clear in the forest fire detection and monitoring applications mentioned in Chap. 8. In the real scenario, piloted airborne means, i.e. airplanes and helicopters, are used today in forest fire activities. Then, the coordination of these aerial means with the unmanned aerial vehicles is a must. In other disaster and security applications, the situation is similar. Then, the methods presented in Chap. 2 should be extended for the integration with the conventional aircrafts.

In general, the lack of integration of the UAVs with the existing air traffic control systems is a main barrier for many commercial applications. This is related to the required certification to fly in civil airspaces. Another barrier is the lack of standard/modular platforms and standardized components, and the development of common UAV interoperability standards. The development of these regulations and standards will play an important role in the practical application of the technologies presented in this book.

Author Index

Alami, Rachid 15

Béjar, Manuel 111

Caballero, Fernando 67
Cuesta, Federico 111, 189

Deeg, Carsten 111

Ferruz, Joaquín 67
Forssén, Per-Erik 67

Gancet, Jérémi 15

Hattenberger, Gautier 15
Hommel, Günter 49, 111

Lacroix, Simon 15, 147
Lemaire, Thomas 15
López-Pichaco, Francisco 189

Martínez-de-Dios, J.R. 207
Maza, Iván 1, 229
Merino, Luis 67, 207
Musial, Marek 49, 111

Nogales, Carlos 189

Ollero, Aníbal 1, 67, 111, 189, 207, 229

Peña, Nicolás 189

Remuß, Volker 49, 111
Ribeiro, Luis M. 207

Solaque, Leonardo 147

Viegas, Xavier 207

Wiklund, Johan 67

Springer Tracts in Advanced Robotics

Edited by B. Siciliano, O. Khatib and F. Groen

Vol. 37: Ollero, A.; Maza, I. (Eds.)
Multiple Heterogeneous Unmanned Aerial Vehicles
233 p. 2007 [978-3-540-73957-9]

Vol. 36: Buehler, M.; Iagnemma, K.; Singh, S. (Eds.)
The 2005 DARPA Grand Challenge – The Great Robot Race
520 p. 2007 [978-3-540-73428-4]

Vol. 35: Laugier, C.; Chatila, R. (Eds.)
Autonomous Navigation in Dynamic Environments
169 p. 2007 [978-3-540-73421-5]

Vol. 34: Wisse, M.; van der Linde, R.Q.
Delft Pneumatic Bipeds
136 p. 2007 [978-3-540-72807-8]

Vol. 33: Kong, X.; Gosselin, C.
Type Synthesis of Parallel Mechanisms
272 p. 2007 [978-3-540-71989-2]

Vol. 32: Milutinović, D.; Lima, P.
Cells and Robots – Modeling and Control of Large-Size Agent Populations
130 p. 2007 [978-3-540-71981-6]

Vol. 31: Ferre, M.; Buss, M.; Aracil, R.; Melchiorri, C.; Balaguer C. (Eds.)
Advances in Telerobotics
500 p. 2007 [978-3-540-71363-0]

Vol. 30: Brugali, D. (Ed.)
Software Engineering for Experimental Robotics
490 p. 2007 [978-3-540-68949-2]

Vol. 29: Secchi, C.; Stramigioli, S.; Fantuzzi, C.
Control of Interactive Robotic Interfaces – A Port-Hamiltonian Approach
225 p. 2007 [978-3-540-49712-7]

Vol. 28: Thrun, S.; Brooks, R.; Durrant-Whyte, H. (Eds.)
Robotics Research – Results of the 12th International Symposium ISRR
602 p. 2007 [978-3-540-48110-2]

Vol. 27: Montemerlo, M.; Thrun, S.
FastSLAM – A Scalable Method for the Simultaneous Localization and Mapping Problem in Robotics
120 p. 2007 [978-3-540-46399-3]

Vol. 26: Taylor, G.; Kleeman, L.
Visual Perception and Robotic Manipulation – 3D Object Recognition, Tracking and Hand-Eye Coordination
218 p. 2007 [978-3-540-33454-5]

Vol. 25: Corke, P.; Sukkarieh, S. (Eds.)
Field and Service Robotics – Results of the 5th International Conference
580 p. 2006 [978-3-540-33452-1]

Vol. 24: Yuta, S.; Asama, H.; Thrun, S.; Prassler, E.; Tsubouchi, T. (Eds.)
Field and Service Robotics – Recent Advances in Research and Applications
550 p. 2006 [978-3-540-32801-8]

Vol. 23: Andrade-Cetto, J,; Sanfeliu, A.
Environment Learning for Indoor Mobile Robots – A Stochastic State Estimation Approach to Simultaneous Localization and Map Building
130 p. 2006 [978-3-540-32795-0]

Vol. 22: Christensen, H.I. (Ed.)
European Robotics Symposium 2006
209 p. 2006 [978-3-540-32688-5]

Vol. 21: Ang Jr., H.; Khatib, O. (Eds.)
Experimental Robotics IX – The 9th International Symposium on Experimental Robotics
618 p. 2006 [978-3-540-28816-9]

Vol. 20: Xu, Y.; Ou, Y.
Control of Single Wheel Robots
188 p. 2005 [978-3-540-28184-9]

Vol. 19: Lefebvre, T.; Bruyninckx, H.; De Schutter, J. Nonlinear Kalman Filtering for Force-Controlled Robot Tasks
280 p. 2005 [978-3-540-28023-1]

Vol. 18: Barbagli, F.; Prattichizzo, D.; Salisbury, K. (Eds.)
Multi-point Interaction with Real and Virtual Objects
281 p. 2005 [978-3-540-26036-3]

Vol. 17: Erdmann, M.; Hsu, D.; Overmars, M.; van der Stappen, F.A (Eds.)
Algorithmic Foundations of Robotics VI
472 p. 2005 [978-3-540-25728-8]

Vol. 16: Cuesta, F.; Ollero, A.
Intelligent Mobile Robot Navigation
224 p. 2005 [978-3-540-23956-7]

Vol. 15: Dario, P.; Chatila R. (Eds.)
Robotics Research – The Eleventh International Symposium
595 p. 2005 [978-3-540-23214-8]

Vol. 14: Prassler, E.; Lawitzky, G.; Stopp, A.; Grunwald, G.; Hägele, M.; Dillmann, R.; Iossifidis. I. (Eds.)
Advances in Human-Robot Interaction
414 p. 2005 [978-3-540-23211-7]

Vol. 13: Chung, W.
Nonholonomic Manipulators
115 p. 2004 [978-3-540-22108-1]

Vol. 12: Iagnemma K.; Dubowsky, S.
Mobile Robots in Rough Terrain – Estimation, Motion Planning, and Control with Application to Planetary Rovers
123 p. 2004 [978-3-540-21968-2]

Vol. 11: Kim, J.-H.; Kim, D.-H.; Kim, Y.-J.; Seow, K.-T.
Soccer Robotics
353 p. 2004 [978-3-540-21859-3]

Vol. 10: Siciliano, B.; De Luca, A.; Melchiorri, C.; Casalino, G. (Eds.)
Advances in Control of Articulated and Mobile Robots
259 p. 2004 [978-3-540-20783-2]

Vol. 9: Yamane, K.
Simulating and Generating Motions of Human Figures
176 p. 2004 [978-3-540-20317-9]

Vol. 8: Baeten, J.; De Schutter, J.
Integrated Visual Servoing and Force Control – The Task Frame Approach
198 p. 2004 [978-3-540-40475-0]

Vol. 7: Boissonnat, J.-D.; Burdick, J.; Goldberg, K.; Hutchinson, S. (Eds.)
Algorithmic Foundations of Robotics V
577 p. 2004 [978-3-540-40476-7]

Vol. 6: Jarvis, R.A.; Zelinsky, A. (Eds.)
Robotics Research – The Tenth International Symposium
580 p. 2003 [978-3-540-00550-6]

Vol. 5: Siciliano, B.; Dario, P. (Eds.)
Experimental Robotics VIII – Proceedings of the 8th International Symposium ISER02
685 p. 2003 [978-3-540-00305-2]

Vol. 4: Bicchi, A.; Christensen, H.I.; Prattichizzo, D. (Eds.)
Control Problems in Robotics
296 p. 2003 [978-3-540-00251-2]

Vol. 3: Natale, C.
Interaction Control of Robot Manipulators – Six-degrees-of-freedom Tasks
120 p. 2003 [978-3-540-00159-1]

Vol. 2: Antonelli, G.
Underwater Robots – Motion and Force Control of Vehicle-Manipulator Systems
268 p. 2006 [978-3-540-31752-4]

Vol. 1: Caccavale, F.; Villani, L. (Eds.)
Fault Diagnosis and Fault Tolerance for Mechatronic Systems – Recent Advances
191 p. 2003 [978-3-540-44159-5]

Printing: Mercedes-Druck, Berlin
Binding: Stein+Lehmann, Berlin

MIX
Papier aus verantwortungsvollen Quellen
Paper from responsible sources
FSC® C105338

If you have any concerns about our products,
you can contact us on
ProductSafety@springernature.com

In case Publisher is established outside the EU,
the EU authorized representative is:
Springer Nature Customer Service Center GmbH
Europaplatz 3, 69115 Heidelberg, Germany

Printed by Libri Plureos GmbH
in Hamburg, Germany